NONTHERMAL PRESERVATION OF FOODS

FOOD SCIENCE AND TECHNOLOGY

A Series of Monographs, Textbooks, and Reference Books

1. Flavor Research: Principles and Techniques, *R. Teranishi, I. Hornstein, P. Issenberg, and E. L. Wick*
2. Principles of Enzymology for the Food Sciences, *John R. Whitaker*
3. Low-Temperature Preservation of Foods and Living Matter, *Owen R. Fennema, William D. Powrie, and Elmer H. Marth*
4. Principles of Food Science
   Part I: Food Chemistry, *edited by Owen R. Fennema*
   Part II: Physical Methods of Food Preservation, *Marcus Karel, Owen R. Fennema, and Daryl B. Lund*
5. Food Emulsions, *edited by Stig E. Friberg*
6. Nutritional and Safety Aspects of Food Processing, *edited by Steven R. Tannenbaum*
7. Flavor Research: Recent Advances, *edited by R. Teranishi, Robert A. Flath, and Hiroshi Sugisawa*
8. Computer-Aided Techniques in Food Technology, *edited by Israel Saguy*
9. Handbook of Tropical Foods, *edited by Harvey T. Chan*
10. Antimicrobials in Foods, *edited by Alfred Larry Branen and P. Michael Davidson*
11. Food Constituents and Food Residues: Their Chromatographic Determination, *edited by James F. Lawrence*
12. Aspartame: Physiology and Biochemistry, *edited by Lewis D. Stegink and L. J. Filer, Jr.*
13. Handbook of Vitamins: Nutritional, Biochemical, and Clinical Aspects, *edited by Lawrence J. Machlin*
14. Starch Conversion Technology, *edited by G. M. A. van Beynum and J. A. Roels*
15. Food Chemistry: Second Edition, Revised and Expanded, *edited by Owen R. Fennema*

NONTHERMAL PRESERVATION OF FOODS

GUSTAVO V. BARBOSA-CÁNOVAS
USHA R. POTHAKAMURY
ENRIQUE PALOU

Biological Systems Engineering Department
Washington State University
Pullman, Washington

BARRY G. SWANSON

Food Science and Human Nutrition Department
Washington State University
Pullman, Washington

Marcel Dekker, Inc.   New York · Basel · Hong Kong

Library of Congress Cataloging-in-Publication Data

Nonthermal preservation of foods / Gustavo V. Barbosa-Cánovas... [et al.].
    p.   cm. -- (Food science and technology ; v. 82)
    Includes bibliographical references and index.
    ISBN 0-8247-9979-8 (hardcover : alk. paper)
    1. Food--Processing. I. Barbosa-Cánovas, Gustavo V.   II. Series: Food
science and technology (Marcel Dekker, Inc.) ; 82.
    TP371.2.N66   1997
    644' .028--dc21   97-35932
                                                                        CIP

The publisher offers discounts on this book when ordered in bulk quantities. For more information, write to Special Sales/Professional Marketing at the address below.

This book is printed on acid-free paper.

MARCEL DEKKER, INC.
270 Madison Avenue, New York, New York 10016
http://www.dekker.com

Current printing (last digit):
10 9 8 7 6 5 4 3 2 1

PRINTED IN THE UNITED STATES OF AMERICA

To our families

Preface

This is the first book specially dedicated to the nonthermal preservation of foods. Most of the literature in the field of nonthermal processing is available in the form of journal articles or patents and usually discusses one or more techniques. Although there are excellent reviews, space availability has limited the amount of detailed information on the different techniques. It was with this concern in mind that the idea of this book was conceived.

Replacing or complementing thermal processes to offer better foods to consumers at a low cost is an interesting proposition. Through the years, many research groups have tried many different approaches to make this proposition a reality. Nonthermal processing technologies are very attractive because of the low processing temperature employed. Foods can be processed at ambient temperatures, thus eliminating the effects of high temperatures on foods. Because of the low temperatures employed in nonthermal processing, the whole process becomes energy-efficient. Although the initial cost for ncessary equipment seems large, the process will prove economical in the long run.

This book presents a thorough and updated review on several nonthermal technologies for preserving foods. It includes experimental results, the mechanisms of microbial inactivation, discussion of sterilization standards, and the applicability of nonthermal processing techniques on a commercial scale. The nonthermal technologies presented include high hydrostatic pressure, pulsed electric fields, oscillating magnetic fields, light pulses, irradiation, and the use of chemicals and bacteriocins as preservation aids. The final chapter is dedicated to hurdle technologies, (i.e., the use of multiple preservation factors that will synergistically interact to favor the preservation of foods). Nonthermal processing can also be applied in combination with conventional methods of preservation, including thermal processes. This final chapter conveys the future in food technology: the effective combination of selected technologies for quality foods at a convenient price.

We sincerely hope this book will be a valuable addition to the food literature and will promote additional research into nonthermal preservation of food.

We would like to thank Washington State University (WSU) for providing an ideal framework in which to develop this book, especially all the faculty and staff members of the Center for Nonthermal Processing of Food, all visiting faculty who come to WSU to work on nonthermal processing, and all the WSU research associates, graduates, and undergraduates involved in the development of emerging technologies. Their contributions to our program on nonthermal processing have been decisive; without them this book would have not been possible.

We express our sincere gratitude to the Honorable Thomas S. Foley, former Speaker of the House; the late Patrick Ormsby, congressional aide to former Speaker Foley; and his staff for their vision, continuous support, and belief in the future of food technology.

A special thanks goes to the Bonneville Power Administration (BPA), Walla Walla, Washington, for believing and investing in our dream to develop nonthermal processing techniques. BPA has contributed significantly to the development of pulsed electric fields and high hydrostatic pressure for the preservation of foods and to the establishment of the Center for Nonthermal Processing of Food. It is a real pleasure to work with an institution that is committed to the well-being of the community, seeks cooperation with other institutions with common goals, and fosters the best possible use of our valuable energy resources. Our special gratitude is offered to Tom Osborn and Jennifer Eskil, who are excellent examples of how to work in partnership with other organizations for a better tomorrow.

We are appreciative of the WSU IMPACT Center for its continuous support in the development of two very important technologies for the future: pulsed electric fields and high hydrostatic pressure. The effort of Dr. A. Desmond O'Rourke, IMPACT Center Director, to promote and sponsor these technologies has been extremely valuable in the promotion of nonthermal processing of foods.

We all thank Universidad de las Américas-Puebla, Mexico; the Institute of International Education (IIE); and the Fulbright/CONACyT (Mexico) Program for supporting Enrique Palou's doctoral studies at WSU.

Our gratitude is also extended to Ms. Sandy Jamison and Ms. Dora Rollins, Center for Nonthermal Processing of Food, for their invaluable editorial assistance.

Gustavo V. Barbosa-Cánovas
Usha R. Pothakamury
Enrique Palou
Barry G. Swanson

Contents

1

Emerging Technologies in Food Preservation

Nonthermal methods for the preservation of foods are under intense research to evaluate their potential as an alternative or complementary process to traditional methods of food preservation. Traditionally, most preserved foods are thermally processed by subjecting the food to a temperature of 60° to 100°C for a few seconds to minutes (1). During this period, a large amount of energy is transferred to the food. This energy may trigger unwanted reactions in the food, leading to undesirable changes or formation of by-products. For example, thermally processed milk may have a cooked flavor accompanied by a loss of vitamins, essential nutrients, and flavors. The fact that not only the shelf life but also the quality of food is important to consumers gave birth to the concept of preserving foods using nonthermal methods. Nonthermal methods of food preservation are being developed to eliminate—or at least minimize—the quality degradation of foods that results from thermal processing.

During nonthermal processing, the temperature of the food is held below the temperatures normally used in thermal processing. Therefore, the quality degradation expected from high temperatures is minimal in nonthermal processing. The vitamins, essential nutrients, and flavors are expected to undergo minimal or no changes during nonthermal processing. Also, nonthermal processes utilize less energy than thermal processes.

Foods can be processed nonthermally using high hydrostatic pressure, oscillating magnetic fields, high intensity pulsed electric fields, intense light pulses, irradiation, chemicals, biochemicals, and hurdle technology. Although these technologies have been used for a long time to inactivate microorganisms and/or preserve food, they have gained recognition as nonthermal methods of food preservation only in the recent past.

The effects of high pressure on microorganisms present in milk, meats, vegetables, and fruits were studied by Hite (2) as early as 1899. Application of a pressure of 6800 atm on grape juice for 10 min stopped fermentation. Peaches and pears subjected to a pressure of 4100 atm for 30 min exhibited a shelf life of 5 years. However, attempts to preserve tomatoes, peas, beans, and beets using high pressure were not successful (2). Getchell (3) described a process for pasteurization of milk using electricity in 1935. Milk processed with electricity tasted better than thermally processed milk. However, inactivation of microorganisms resulted from the heat generated by the electricity. Kimball (4) studied the effects of magnetic fields on the growth and reproduction of bacteria and yeasts. The budding of yeasts was inhibited by heterogeneous magnetic fields. The use of chemicals such as salt, sugar, and/or vinegar is the oldest known method of food preservation. Irradiation and the concept of hurdle technology for food preservation have also been reported for many years; use of pulsed high intensity light to inactivate microorganisms is a relatively new technology.

The application of high hydrostatic pressure for food processing consists of subjecting the food to pressures in the range of 4000 to 9000 atm (5). High hydrostatic pressure is used for the inactivation of microorganisms and certain enzymes and for shelf-life extension of the foods. Spores can be inactivated by combining high pressure with high temperature. The germination of spores is an important step in spore inactivation. High pressure promotes the germination of spores, and high temperature inactivates the germinated spores. Some of the applications of high pressure technology include modification of the texture and sensory properties of foods, tenderization of pre-rigor beef, gelation of surimi, manufacture of food purees, jams and jellies from strawberries, marmalade from oranges, and shelf-life extension of milk. One of the important challenges in using high pressure technology is the fabrication of pressure vessels and seals that can withstand the high pressures during the cycles of pressurization and depressurization.

Oscillating magnetic fields with a magnetic flux of 5–50 tesla (T) and frequency of 5–500 kHz are reported (6) to cause inactivation of microorganisms. Foods with an electrical resistivity of 10–25 ohms-cm may be sealed in a plastic bag and subjected to oscillating magnetic fields. The fermentation of foods such as cheese, beer, or yogurt may be stopped with magnetic fields. One of the attractive features of using magnetic fields for food preservation is that the food can be packaged prior to processing, reducing the possibility of cross-contamination during packaging. However, only flexible film or paper packaging material can be used; reflective metallic packaging cannot be used (6).

High intensity electric fields applied to a food in the form of short-duration pulses can inactivate the microorganisms and certain enzymes (7). Spores can be inactivated by combining methods such as pulsed electric fields with temperature and lysozyme. High intensity pulsed electric field processing systems include a power source; capacitor bank; switch; treatment chamber; aseptic packaging apparatus; and voltage, current, and temperature probes. A properly designed treatment chamber with uniform electric field distribution and no field enhancement points is important for the electric field process. The treatment chamber may be designed to operate in a static or continuous mode. In the static chamber, each volume of the food is treated separately. In the continuous and coaxial chambers, the food flows continuously while being processed.

Pulsed electric field induced microbial inactivation is dependent on several factors, the most important being electric field intensity, treatment time, temperature of the food, and type of microorganism. Microbial inactivation increases with an increase in the electric field intensity, exposure time, and temperature of the food (8,9). However, it is desirable to maintain the temperature below 30–40°C by providing a cooling system. Different bacteria have different sensitivities to electric field treatment. In general, gram-positive bacteria and yeasts are more resistant to electric fields than gram-negative bacteria (10). The optimum conditions for maximum inactivation of a specific microorganism can be determined after preliminary research.

Application of light pulses involves exposure of foods to short-duration pulses (1 μs to 0.1 s) of intense incoherent light. Light with an energy density of about 0.01 to 50 J/cm^2 and wavelength in the range of 170 to 2600 nm is used. Such incoherent intense pulses of light may be generated using gas-filled flashlamps or spark gap discharge apparatus. Full- or filtered-spectrum light may be used, depending on the degree of sterilization expected. The filtered-spectrum light is devoid of wavelengths known to cause undesirable reactions in foods. Glass or liquid filters are used to obtain the filtered-spectrum. In general, filtered light is more effective for microbial inactivation than full-spectrum light. Pulsed light technology is applicable primarily for surface pasteurization of meats, fish, vegetables, fruits, bakery and solid milk products, bulk sterilization of transparent homogeneous fluids, and sterilization of the packaging material used in aseptic processing (11).

Irradiation of foods was one of the earliest nonthermal food preservation technologies. Irradiation is the exposure of food to radiation with energy of 5 to 10 kGy and wavelengths of 2000 Å or less (1). Ultraviolet, beta, gamma, and x-rays and microwaves are included in this wavelength range. One of the attractions of irradiation is its capability of pasteurizing foods in the frozen state (12). The World Health Organization (WHO) approved a radiation dosage up to 10 kGy as being "unconditionally safe for human consumption" (13). Radiation is approved for food applications in at least 43 countries, including the United States. However, one of the problems in getting radiation approved in the United States for broader applications is the US Food and Drug Administration definition of irradiation. Irradiation is considered to be a food additive and not a process. Therefore, irradiated food must be labeled (1). Some of the applications of irradiation include extension of the shelf life of root crops; disinfestation of spices, fruits, and cereals; delay of fruit ripening; improvement of the sensory properties of foods; and sterilization of containers in aseptic packaging. Irradiation has the potential to replace the use of many hazardous chemical pesticides and preservatives.

Hurdles are physical or chemical parameters that can be adjusted to ensure the microbial stability and safety of foods. The physical parameters include processing and storage temperatures, water activity, pH, redox potential of the food, and preservatives used in the food. The parameters are controlled at levels that inhibit or inactivate the microorganisms and thus render the food safe (14). Hurdle technology is used in the preservation of meat and seasonal or regional fruits and vegetables.

Chemical and biochemical hurdles include antimicrobials, antioxidants, humectants, sequestrants, and bacteriocins. Antimicrobials may be naturally present in foods, formed in response to a stress, or added for the purpose of preservation. Enzymes such as lactoperoxidase, proteins such as casein and lactoferrin, or pigments such as anthocyanin present in foods possess antimicrobial properties. Organic acids, such as citric, malic, sorbic, tartaric, propionic, and benzoic are potential acidulants added to foods to prolong shelf life (15,16). Bacteriocins are protein-containing molecules with the ability to inhibit food-borne pathogens. Colicins produced by *Escherichia coli* and nisin produced by lactic acid bacteria are naturally produced bacteriocins (17). Nisin is used to inhibit gassing caused by *Clostridium* in cheese and spore-forming gram-positive bacteria, gram-negative bacteria, and yeasts in cheese spreads, canned vegetables, soups, and puddings. Addition of nisin to chocolate milk and evaporated milk reduces the thermal processing time (18).

High hydrostatic pressure and hurdle technology are being used on a commercial scale. The potential applications of pulsed electric fields, oscillating magnetic fields, and light pulses for the stabilization of foods are still be-ing studied.

Each of the nonthermal technologies has specific applications in terms of the types of food that can be processed. For example, high pressure, oscil-lating magnetic fields, antimicrobials, light pulses, and hurdle technology are useful in processing both liquid and solid foods, whereas pulsed electric fields are more suitable for liquid foods and irradiation is useful for solid foods. Furthermore, light pulses are more useful for surface pasteurization. In addi-tion, light pulses, irradiation, and magnetic fields can be used to process prepackaged foods, reducing the risk of cross- or post-process contamination. Therefore, nonthermal technologies are not applicable to the processing of all kinds of foods. Each nonthermal technology has its merits and limitations. In many cases—for example, in the inactivation of spores—it is necessary to use a combined methods approach.

Although preliminary studies indicate promising results, it will be a while before most nonthermal technologies can be used on a commercial scale. Most research is focused on applications for nonthermal technologies, and little on understanding the mechanisms of inactivation of microorgan-isms, spores, or enzymes. The ultrastructural changes in pulsed electric field–treated cells of *Saccharomyces cerevisiae*, *E. coli*, and *Staphylococcus aureus* and pressure-treated cells of *S. cerevisiae* were observed using elec-tron microscopy in an attempt to understand the inactivation mechanisms. Electric field treatment resulted in cell membrane rupture in cells of *S. cere-visiae* and *S. aureus*. The cell membrane in *E. coli* cells was not obviously damaged by the electric field. High pressure treatment results in membrane disruption of *S. cerevisiae* cells. These studies indicate cell membrane dam-age or rupture that may or may not be the primary inactivation mechanism. Therefore, additional studies are necessary to arrive at the inactivation mech-anism. With a proper understanding of the inactivation mechanisms of non-thermal technologies, some of the limitations can potentially be overcome —or where one technology fails, others can be used.

Nonthermal processes are expected to induce only minimum quality degradation of food. It is therefore necessary to evaluate changes in sensory attributes, concentrations, and structures of lipids, proteins, carbohydrates, and other constituents of foods. It is also important to compare the quality and shelf life of foods treated by different nonthermal methods to determine if one method is more suitable than others for a specific food.

The most important issue involved in the commercialization of nonther-mal technologies is regulatory approval. Foods processed thermally or nonthermally must comply with the safety regulations set forth by the Food and Drug Administration prior to being marketed or consumed. For example, the regulations for thermally processed low-acid canned foods are contained in Title 21, Part 113 of the US Code of Federal Regulations, entitled "Thermally Processed Low-Acid Foods Packaged in Hermetically Sealed Containers." The regulations in Title 21 were established to evaluate (a) adequacy of the equipment and procedures to perform safe processing operations, (b) adequacy of record keeping to prove safe operation, (c) justification of the adequacy of process time and temperature used, and (d) qualifications of supervisory staff responsible for thermal processing and container closure operations (19).

It is not enough, therefore, to develop a food processing method that satisfies only pasteurization or sterilization requirements; it is necessary to ensure that the process is safe for the equipment operators and consumers. For example, there are health concerns regarding the effects of magnetic fields on operators. Until it can be assured there are no health risks in operating the magnetic field equipment, or safety measures are provided to avoid such health risks, it will not be possible to commercialize the process. Necessary safety measures must be specified for the operation of high voltage and high pressure equipment.

The two types of bacteria of concern in thermal food preservation are organisms of public health significance and spoilage-causing bacteria. In low-acid foods with a pH greater than 4.6, the organism of public health significance is *Clostridium botulinum*. Canned foods are processed based on the survival probability for *C. botulinum* of 10^{-12} or one survivor in 10^{12} cans. The organism most frequently used to characterize low acid food spoilage by mesophilic spore-formers is PA 3679, a strain of *C. sporogenes*. Most food companies accept thermal inactivation of 10^{-5} for mesophilic spore-formers and 10^{-2} for thermophilic spore-formers. The processing time depends on the bioburden of the most resistant bacteria in a particular food, the spoilage risk involved, and whether the food can support the growth of potential contaminating bacteria (19).

Research demonstrates that bacteria, yeasts, and molds are not equally susceptible to nonthermal methods. The question to address is whether the processing time of nonthermal processed foods is based on the most resistant organism or on chemical constituents present in the food. Also, do microorganisms follow a linear and predictable inactivation course during nonthermal processing? For example, is a specific organism that is resistant to electric fields also resistant to high pressure?

The validation of each nonthermal method and determination of compliance regulations necessary for commercialization are complex and challenging. Progress in validation is encouraging. Regulatory compliance will be addressed in the near future.

REFERENCES

1. J. M. Jay, *Modern Food Microbiology*, 4th ed., Van Nostrand Reinhold, New York, 1992.
2. D. G. Hoover, C. Metrick, A. M. Papineau, D. F. Farkas, and D. Knorr, Biological effects of high hydrostatic pressure on food microorganisms. *Food Technol.* 43: 99–107 (1989).
3. B. E. Getchell, Electric pasteurization of milk. *Agric. Eng.* 16: 408–410 (1935).
4. G. C. Kimball, The growth of yeast in a magnetic field. *J. Bacteriol.* 35: 109–122 (1938).
5. U. R. Pothakamury, G. V. Barbosa-Cánovas, B. G. Swanson, and R. S. Meyer, The pressure builds for better food processing. *Chem. Eng. Prog.*, March pp. 45–55 (1995).
6. G. A. Hofmann, Deactivation of microorganisms by an oscillating magnetic field. U. S. Pat. 4,524,079 (1985).
7. G. V. Barbosa-Cánovas, U. R. Pothakamury, and B. G. Swanson, State of the art technologies for the stabilization of foods by nonthermal processes: physical methods *Food Preservation by Moisture Control. Fundamentals and Applications.* (G. V. Barbosa-Cánovas and J. Welti-Chanes, eds.), Technomic, Lancanster, PA, 1995, pp. 493-532.
8. B. Qin, U. R. Pothakamury, G. V. Barbosa-Cánovas, and Swanson, Nonthermal pasteurization of liquid foods using high intensity pulsed electric fields. *Crit. Rev. Food Sci. Nutrition* 36(6): 603-627 (1996).
9. B. Qin, H. Vega-Mercado, U. R. Pothakamury, G. V. Barbosa-Cánovas, and B. G. Swanson, Application of pulsed electric fields for inactivation of bacteria and enzymes. *J. Franklin Inst.* 332A: 209-220 (1995).
10. H. Hülsheger, J. Potel, and E. G. Niemann, Killing of bacteria with electric pulses of high field strength. *Radiat. Environ. Biophys.* 20: 53-65 (1981).
11. J. E. Dunn, R. W. Clark, J. F. Asmus, J. S. Pearlman, K. Boyer, and F. Pairchaud, Methods and apparatus for preservation of foodstuffs. Int. Pat. Appl. No. WO 88/03369 (1988).
12. M. K. Wagner and L. J. Moberg, Present and future use of traditional antimicrobials. *Food Technol.* 43(1): 133 (1989).
13. S. Thorne, *Food Irradiation*, Elsevier Appl. Sci., London, 1991.
14. L. Leistner, Introduction to hurdle technology, *Food preservation by combined processes*, Final Report of FLAIR concerted action No. 7, subgroup B (L. Leistner and L. G. M. Morris, eds.) , 1994, pp. 1–6.

15. L. R. Beuchat and D. A. Golden, Antimicrobials occurring naturally in foods. *Food Technol.* 43(1): 134-142 (1989).
16. P. M. Davidson and A. L. Branen, *Antimicrobials in Foods,* Marcel Dekker, New York, 1993.
17. P. Muriana, Potential applications of bacteriocins in processed foods, *Advances in Aseptic Processing Technologies* (R. K. Singh and P. E. Nelson, eds.), Elsevier Appl. Sci. London, 1992.
18. M. A. Daeschel, Antimicrobial substances from lactic acid bacteria for use as food preservatives. *Food Technol.* 43 (1): 164-167 (1989).
19. A. Teixeira, Thermal process calculations, *Handbook of Food Engineering* (D. R. Heldman and D. B. Lund, eds.), Marcel Dekker, New York, 1992.

2

High Hydrostatic Pressure Food Processing

I. INTRODUCTION

Food can be preserved by thermal or nonthermal processes. The thermal processes—namely, heating and cooking—affect the quality of the food in addition to inactivating microorganisms. On the other hand, nonthermal processes can be used for the inactivation of food-spoilage microorganisms without affecting the quality of the food. High pressure technology is gaining importance in the food industry because of the advantages of inactivating the microorganisms and enzymes and producing high quality foods (1).

High pressure technology was originally used in the production of ceramics, steels, and superalloys. In the past decade, high pressure technology expanded to the food industry. High pressure inactivation of microorganisms has been recognized since the beginning of the twentieth century, but only in the past decade did researchers begin to study potential commercialization of high pressure technology in the food industry. At a pressure of 4000–9000 atm, enzymes and bacteria are inactivated, but taste and flavor remain unaffected. Because pressure is uniform throughout the food, preservation of food is uniform, with no part escaping preservation. Unlike thermal treatment, pressure treatment is not time/mass dependent, thus reducing the processing time (2).

II. ENGINEERING ASPECTS OF HIGH PRESSURE TECHNOLOGY

High isostatic pressure technology is the application of pressure uniformly throughout a product, and it is basically applied for isostatic pressing, quartz growing, chemical reactors, and simulators (3). Quartz crystals are grown from a strong alkaline solution of sodium hydroxide at a pressure of up to 2000 atm and a temperature of up to 420°C. Some chemical reactions are carried out at high pressure to increase the yield of the reaction. For example, low-density polyethylene is synthesized at a pressure of 2000 atm and a temperature of 350°C. High pressure vessels are also used as simulators to test equipment that would be used in a high pressure environment; for example, in deep sea. The food industry employs the technique of isostatic pressing for applying high pressures to foods.

A. Generation of High Pressure

Pressure vessels that can withstand high pressures are in use in the metal and ceramic industries. However, the food industry demands equipment that can tolerate much higher pressures of about 4000 atm with a more efficient and durable cycling of about 100,000 cycles/year. High pressures can be generated by the following (3):

1. *Direct compression* is generated by pressurizing a medium with the small-diameter end of a piston (Fig. 1). The large-diameter end of the piston is driven by a low pressure pump. This direct compression method allows very fast compression, but the limitations of the high pressure dynamic seal between the piston and the vessel internal surface restricts the use of this method to small-diameter laboratory or pilot plant systems.
2. *Indirect compression* uses a high pressure intensifier to pump a pressure medium from a reservoir into a closed high pressure vessel until the desired pressure is reached (Fig. 2). Most industrial isostatic pressing systems utilize the indirect compression method.
3. *Heating of the pressure medium* utilizes expansion of the pressure medium with increasing temperature to generate high pressure. Heating of the pressure medium is therefore used when high pressure is applied in combination with high temperature; it requires very accurate temperature control within the entire internal volume of the pressure vessel.

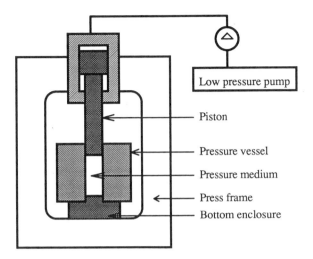

FIGURE 1 Generation of high pressure by direct compression of the pressure transmitting medium. (From Ref. 4.)

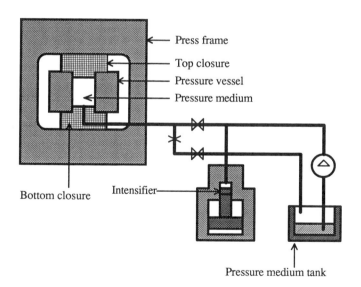

FIGURE 2 Generation of high pressure by indirect compression of the pressure transmitting medium. (From Ref. 4.)

The isostatic pressing systems may be operated as cold isostatic, warm isostatic, or hot isostatic systems (4):

1. *Cold isostatic pressing* (CIP) is essentially a forming technique used in the metal, ceramics, carbon/graphite, and plastic industries. Powdered materials are filled in an elastomer mold and subjected to high pressure. Pressure employed is in the range of 500–6000 atm. The CIP process uses "wet bag" or "dry bag" configurations. In the wet bag method, the mold is filled outside of the pressure vessel (Fig. 3). The mold is then placed in the pressure vessel, which is filled with the pressure medium. In the dry bag method, the mold is fixed in the pressure vessel and separated from the pressure medium by an elastomer tool. The cycle time in a wet bag method is a few minutes; the

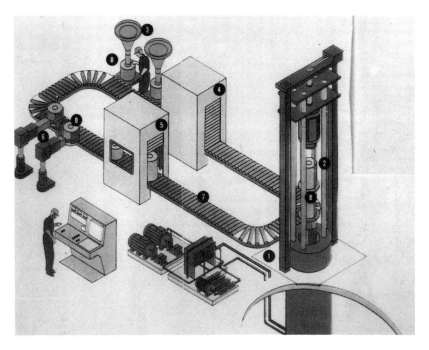

FIGURE 3   A wet bag isostatic plant. 1) Automated wet bag isostatic press, 2) equipment for loading mould into pressure vessel, 3) powder dosing and mould filling equipment, 4) wash booth, 5) drying booth, 6) compact removal unit, 7) tooling conveyor, 8) tooling. (Courtesy of Engineered Pressure Systems, Inc.)

cycle time in a dry bag method varies between 20 and 60 s. CIP is the most promising technique for use in the food industry.

2. *Warm isostatic pressing* (WIP) is also a forming technique. Isostatic pressure is applied in combination with temperatures between ambient and 200°C. WIP is used in situations in which a chemical reaction develops during pressurization.

3. *Hot isostatic pressing* (HIP) is used primarily in metallic and ceramic industries. The material is uniformly heated and pressurized. The temperature employed is as high as 2200°C and pressure is 1000–4000 atm. The pressure medium used is a gas such as argon, nitrogen, helium, or air. The cycle time typically varies between 6 and 12 h (3).

The CIP equipment originally developed for ceramic application was modified to meet additional requirements for food processing. Although the pressure medium used in the CIP vessel is water containing an anti-rusting agent or synthetic oil to protect the pressure vessel against corrosion, food processing requires use of potable water. The forming pressure used in the ceramic industry is normally 1000 to 3000 atm with a holding time of 10 s to 1 min. Food processing requires holding times ranging between 5 and 20 min at pressures greater than 4000 atm.

B. High Pressure Equipment

The Japanese are the leading manufacturers of high pressure vessels. The major Japanese companies manufacturing high pressure vessels are Mitsubishi Heavy Industries Ltd., Kobe Steel Ltd., and Nippon Steel Ltd. Other manufacturers of high pressure equipment include Engineered Pressure Systems, ABB Autoclave Systems Inc., ACB, NKK Corp., and Autoclave Engineers (5). Figure 4 presents several equipments. The first high pressure food processing vessel was manufactured by Mitsubishi Heavy Industries, Tokyo. Table 1 lists the pressure vessels manufactured by Mitsubishi Heavy Industries (6).

To minimize the reduction of equipment life due to repeated use of the pressure vessel, the Mitsubishi pressure vessel is made up of double cylinders. The inner surface of the pressure vessel is preloaded with a high compression stress. The parts of the vessel that come in contact with the pressurizing medium are made of stainless steel. The pressurizing and depressurizing cycle is fast. Maximum pressure is attained in 90 s. High pressure and long holding time imply that a great load is applied to the seal of the vessel cover. A self-seal packing with high durability and reliability is used (Fig. 5). The seals can withstand repeated opening and closing of the pressure vessel and application

(A)

(B)

FIGURE 4   (A) Laboratory scale pressure vessel. (B) Mobile pressure vessel. (Courtesy of ABB Autoclave Systems, Inc.)

TABLE 1 Specifications of High Pressure Vessels Manufactured by Mitsubishi Heavy Industries

Mitsubishi Model	Diameter (m)	Length (m)	Volume (m³)	Maximum Operating Pressure (atm)
MFP 700	0.06	0.2	0.6	7000
MCT 150	0.15	0.3	6.0	4200
FP-30V	3.0	7.0	50	4200
FP-40L	4.0	17.0	210	4000

Source: From Ref. 6.

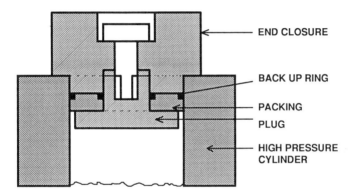

FIGURE 5 A self-seal packing used in high pressure vessels. (From Ref. 7.)

of high pressure without leakage. A piston driven by a hydraulic cylinder is used to generate the required pressure (7).

The productivity of the batch system is increased by a reduction in the pressurizing–depressurizing cycle. The pressurizing time is reduced by increasing the delivery rate of the pump. When the required operating pressure is attained, the pumping rate is reduced. At the end of the specified holding time, the pressure vessel is depressurized in two stages to avoid sudden release of pressurized water (7).

a. Batch and continuous processing equipment

In a batch processing system, food is pressurized in batches, one after the other. Batch processing reduces the risk of large quantities of food becoming contaminated by the lubricants or wear particles from the machinery. Different types of food can be processed in a batch system without the danger of cross-contamination or the need to clean the equipment after each run. The technical advantage of the batch-type pressure vessel is the simplicity of fabrication when compared with a continuous flow pressure vessel operating at pressures as high as 4000–9000 atm (4). A batch system pressure vessel with a processing capacity of 600 L/h of liquid food at a maximum operating pressure of 4200 atm was used to produce grapefruit juice commercially in Japan. The juice had a fresh-like flavor with no bitterness (7).

The production rate of the batch process can be increased by operating pressure vessels in sequence with no lag in the processing times so the system operates sequentially (Fig. 6). A free piston separates the juice from the potable water.

Kobe Steel Ltd. developed a small test pressure vessel in addition to one of the largest pressure vessels available today. The dimensions and internal volume capacities are listed in Table 2. The small test pressure vessel uses a piston for pressurization; the oil hydraulic system and operational panel are compactly packaged as part of the equipment. Operation is fully automated, and the temperature inside the pressure vessel can be recorded. The equipment also allows the use of a pressure control program (8).

The research high pressure food processing system developed by ABB Autoclave Systems, Inc., Vasteras, Sweden, consists of two components: the process module and the control module. The process module consists of a cabinet that contains the Quintus™ prestressed wire-wound pressure vessel, the electrohydraulic pumping system, and a hot water circulation system. The system illustrated in Figure 7 has a chamber size of 90 mm diameter × 225 mm length and can reach 9000 atm at 80°C. The maximum pressure is attained within 4 min. Temperature is maintained by circulating water in channels between the wire winding and the cylinder wall of the pressure vessel.

The programmable control module in the ABB high pressure research vessel monitors and controls the process time, pressure, and temperature. A microprocessor is used to control food loading into the press, press cycling, and downloading to a conveyor (9). The cost of treating foods in 100 L and 500 L systems is approximately 25 cents and 7 cents per batch, respectively (10). The internal capacities and maximum operating pressures in the high pressure systems from ABB Autoclave Systems are listed in Table 3.

Pressure vessels with other dimensions available from ABB Autoclave Systems include

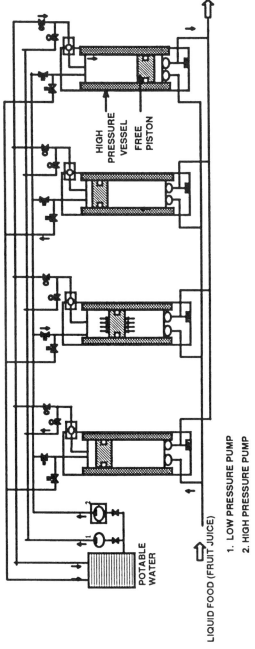

FIGURE 6 High pressure system for processing liquid foods. (From Ref. 7.)

TABLE 2 Specifications of High Pressure Vessels Manufactured by
Kobe Steel Ltd.

Diameter (m)	Length (m)	Volume (liter)	Maximum Operating Pressure (atm)	Maximum Temperature
0.06	0.2	N. A.	7000	80°C
2.0	3.0	9400	1960	80°C

N.A.: not available.
Source: From Ref. 8.

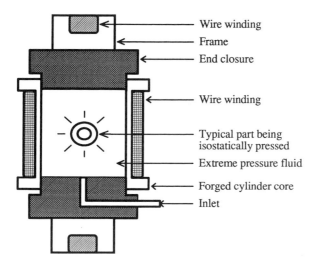

FIGURE 7 Quintus prestressed pressure vessel. (Courtesy of ABB Autoclave Systems, Inc.)

TABLE 3 Specifications of High Pressure Vessels Manufactured by
ABB Autoclave Systems

Model	Diameter (m)	Length (m)	Maximum Operating Pressure (atm)	Volume (liter)	Maximum Temperature
Quintus	0.09	0.225	9,000	N. A.	80°C
Quintus	0.3	1.25	9,000	100	80°C
Quintus	0.5	2.5	9,000	500	80°C

N.A.: not available.
Source: From Ref. 9.

- 0.045 m dia × 0.3 m length with a maximum pressure of 12,000 atm
- 0.11 m dia × 0.26 m length with a maximum pressure of 8300 atm
- 0.32 m dia × 1.25 m length (100 L) with a maximum pressure of 9000 atm
- 0.50 m dia × 2.50 m length (500 L) with a maximum pressure of 9000 atm.

The fatigue value can be maintained infinitely by replacing a shrunk wear-liner every 30,000 cycles. Changing the liner is convenient and inexpensive.

ABB Autoclave Systems is designing and constructing a high pressure vessel to work in batch mode with a maximum operating pressure of 17,000 atm. The internal diameter (ID) of the pressure vessel will be 0.076 m and the internal height (IH) will be 0.18 m (11).

A warm isostatic pressing system is available from the Engineered Pressure Systems Inc. (EPSI), a subsidiary of National Forge Co., Andover, Mass. The system consists of a double-ended, lined pressure vessel with plug closures. The design parameters of the high pressure systems constructed by EPSI are listed in Table 4.

Recently, EPSI developed a laboratory scale pressure vessel with the following specifications: 0.1 m ID × 2.5 m IH with maximum operating pressure of 6800 atm and maximum operating temperature of 90°C. The design pressure is 7500 atm; the maximum operating pressure is 6800 atm. An electrohydraulic intensifier pump with a motor pressurizes the vessel to the operating pressure in 5 min or less (12). An advanced laboratory scale pressure

TABLE 4   Specifications of High Pressure Vessels Manufactured by Engineered Pressure Systems

Diameter (m)	Length (m)	Maximum Operating Pressure (atm)	Volume (liter)
1.7	4.0	1000	9,000
1.0	4.0	2000	3,150
0.6	4.5	4100	1,250
0.6	2.5	5500	7,00
0.25	0.75	6900	37
0.1	1.0	10,300	8.5
0.09	0.55	13,800	3.5

Source: From Ref. 4.

(A)

Fᴵɢᴜʀᴇ 8   (A) Washington State University high hydrostatic pressure system. (B) Washington State University pressure chamber. (Courtesy of Engineered Pressure Systems, Inc.)

vessel with a useful diameter of 0.024 m, length of 0.04 m, and maximum pressure of 8000 atm is also available from EPSI (Fig. 8).

A low maximum operating pressure can cause drastic reductions in the fabrication costs. High pressure processing may be combined with moderately high temperatures, so the operating pressures required are not extremely high (13).

C. Process Description

Operation of a high pressure vessel

A sterilized container filled with food is sealed and placed in the pressure chamber for pressurizing. Ethylene-vinyl alcohol copolymer film (EVOH)

(B)

and polyvinyl alcohol film (PVOH) are recommended for packaging food for high pressure treatment (14). No deformation of the package occurs because the pressure is uniform (10). Once loaded with the food package and closed, the vessel is filled with the pressure-transmitting medium (Fig. 9). In most of the current equipment, the pressure medium used is water mixed with a small amount of soluble oil for lubrication and anticorrosion purposes.

The basis for applying high pressure to foods is to compress the water surrounding the food. At room temperature, the volume of water decreases by 4% at 1000 atm; by 7% at 2000 atm; by 11.5% at 4000 atm; and by 15% at 6000 atm (15). Because liquid compression results in a small volume change,

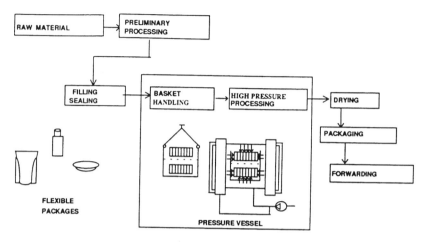

FIGURE 9 High pressure system for processing packaged foods. (From Ref. 7.)

high pressure vessels using water do not present the same operating hazards as vessels using compressed gases (16).

Food is subjected to high pressure for a specified time period. The holding time in the pressure chamber depends on the type of food and process temperature. At the end of the processing time, the chamber is decompressed to remove the treated batch. A new batch of food is placed in the pressure vessel and the cycle begins again (2).

The engineering challenges of the application of high pressure in the food industry are basically the fabrication of pressure vessels to handle large volumes of food and withstand the high pressures. The pressure vessel should have a short cycle time, be easy to clean, and be safe to operate, with accurate process controls. It is desirable to develop a continuous process of pressurization for industrial purposes at a reasonably low capital and operating costs (3). Most of these challenges are being met to some extent, but research is still being conducted to develop high pressure technology and the necessary equipment.

III. BIOLOGICAL EFFECTS OF HIGH PRESSURE

The study of effects of pressure on living organisms is called barobiology. Pressures higher than atmospheric pressure are called hyperbaric. High pressure induces a number of changes in biological systems—morphological, biochemical, and genetic—as well as changes in the cell membrane and cell wall of microorganisms. In general, gram-negative bacteria are inactivated at a lower pressure than the gram-positive bacteria.

A. Effects of High Pressure on Microorganisms

a. Morphological changes

Most bacteria are capable of growth up to pressures of 200–300 atm. Microorganisms that are capable of growth at pressures higher than 400–500 atm are named barophiles. Barophobic organisms grow poorly or do not grow at pressures higher than 300–400 atm. Microorganisms that can grow in the pressure range of 1–500 atm are called eurybaric. Baroduric organisms survive but cannot grow at pressures as high as 500–2000 atm (17).

Filament formation
Filament formation is a very marked change in microorganisms subjected to high pressure. *Escherichia coli* is highly filamentous at pressures in the range of 270–400 atm. At 400 atm, *E. coli* grows to a length of 10–100 μm, compared to the normal length of 1–2 μm when grown at 1 atm. The pressure-induced filaments are single unsegmented cells of normal width (~0.6 μm) (18). At 400 atm, *Vibrio* spp. form filaments five to eight times longer when compared to cells grown at 1 atm. *Bacillus mycoides* forms cells two to three times longer at 270 atm. *Serratia marinorubra* forms filaments as long as 200 μm at 600 atm compared to the normal length of 0.6–1.5 μm at 1 atm. Not only do species differ in their tendency to form filaments at increased pressures, various strains of the same species exhibit marked differences in this respect. The amount of protein produced per unit length of cell is similar at different pressures in *E. coli*. At high pressures, the amount of cellular RNA produced is significantly greater whereas the amount of cellular DNA is markedly less than that produced in cells grown at ambient pressures (17).

Cessation of motility
Most of the motile bacteria are immobilized by prolonged pressurization at 200–400 atm. At 100 atm, *E. coli*, *Vibrio*, and *Pseudomonas* retain flagella, whereas at 400 atm, these microorganisms lose their flagella (17). Loss of motility is reversible in some bacteria (19).

b. Microbial inactivation

Moderately high pressures decrease the rate of growth and reproduction; very high pressures cause inactivation of microorganisms. The threshold pressure for retardation of reproduction and inactivation is dependent on the microorganism and species.

Growth and reproduction of *E. coli* (ATCC 11303) are retarded when these bacteria are subjected to pressures ranging from 100–500 atm. Reproduction, defined as the increase in number of viable cells, is retarded more than growth. Growth may be defined as the increase in the biomass as indicated by the optical density (18).

When *E. coli* is incubated at 200 atm, its growth rate is faster as temperature is increased. For example, the stationary phase is attained in 10–15 h at 30°C; at 40°C, it takes 5–10 h. The lag phase is prolonged at pressures above 400 atm. At 525 atm there is no growth. However, cells are inactivated at lower pressures when temperature is increased (18).

Metrick et al. (20) compared the heat and pressure resistances of *Salmonella typhimurium* 7136 and *S. senftenberg* 775 W. *S. senftenberg* is a heat-resistant species with a D value of 15 min at 57.5°C; *S. typhimurium* has a D value of 3 min at 57.5°C. The pressure inactivation studies were made in a phosphate buffer solution and in a chicken-base baby food (Fig. 10). The inactivation of both microorganisms is greater in buffer than in chicken baby food. The survivor curves at 3400 atm show a stabilization effect (or "tailing") after an initial reduction (Fig. 11). The stabilization effect was attributed to one of two factors: either a small portion of the bacterial population was more resistant than the majority of the population to pressurization or the rate

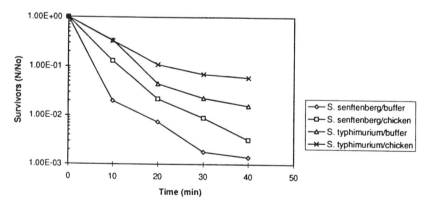

FIGURE 10   Inactivation of *Salmonella* in 63 mM phosphate buffer and strained chicken medium at 2720 atm and 23°C. (From Ref. 20.)

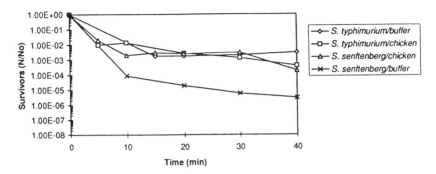

FIGURE 11 Inactivation of *Salmonella* in 63 mM phosphate buffer and strained chicken medium at 3400 atm and 23°C. (From Ref. 20.)

of inactivation was very small. The pressure-resistant population was isolated, grown, and subjected to pressurization. There was no difference in the effects of pressure on the original culture or on the pressure-resistant subculture. When pressure is released, cell recovery occurred in the chicken medium, but not in the buffer. Recovery may be the result of the fact that chicken medium is more nutritious than the buffer solution. One of the key sites of microbial inactivation is the cell membrane. Other possible pressure-sensitive sites that affect viability include the binding of amino acyl-tRNA to ribosome and mRNA, and deactivation of important intracellular enzymes. Presumably, the action of pressure inactivation is similar to thermal inactivation in that more than a single factor is responsible for the death of the bacterial cell.

The inactivation of *Listeria monocytogenes* and *Vibrio parahaemolyticus* was studied by Styles et al. (21). *L. monocytogenes* is a prevalent contaminant in milk and *V. parahaemolyticus* in seafood. Similar to the research of Metrick et al. (20), the inactivation studies were carried out in phosphate buffer and a food medium. The food media used for *L. monocytogenes* was ultra high temperature (UHT) treated milk and raw milk, and that for *V. parahaemolyticus* was sterile heat-processed clam juice. No appreciable decrease in cell population was observed at a pressure of 2380 atm but nearly a 3 log reduction occurred at 2720 atm in the buffer. At 3060 atm, the bacterial population was reduced 3 log cycles in 20 min; at 3400 atm there were fewer than 10 CFU/mL survivors. Though milk is protective against pressure inactivation, treatment at 3400 atm resulted in a 6 log cycle reduction within 80 min in UHT milk and within 60 min in raw milk (Fig. 12).

V. parahaemolyticus was rapidly inactivated by pressures higher than 1700 atm. A population of 10^6 CFU/mL in a buffer was reduced by 6 log cycles within 30 min at 1700 atm (Fig. 13). A 6 log cycle reduction was obtained in clam juice within 10 min at 1700 atm, within 30 min at 1360 atm, and within 40 min at 1020 atm (21) (Fig. 14). Table 5 gives observed pressure inactivation times of selected bacteria.

The baro-sensitivity of microorganisms increases in the order of gram-positive bacteria, yeasts, and gram-negative bacteria. Pork slurries inoculated with *Bacillus cereus, Campylobacter jejuni, Candida utilis, E. coli, Micrococcus luteus, Pseudomonas aeroginosa, Saccharomyces cerevisiae, Salmonella typhimurium, Staphylococcus aureus, Streptococcus faecalis,* and *Yersinia enterocolitica* were subjected to pressures of 3000–6000 atm. The microorganisms, except for the spore-former *B. cereus,* were inactivated, giving counts of less than 10 CFU/g (Figs. 15a,b,c) (22). Coagulation and discoloration were observed in the pressurized slurries. The coagulants were whiter and harder after treatment at

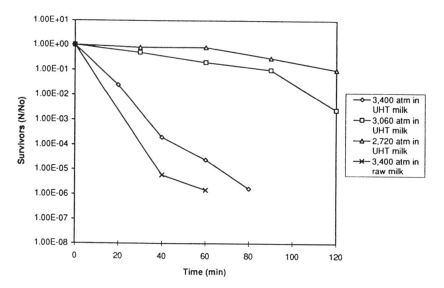

FIGURE 12   Inactivation of *L. monocytogenes* in milk. (From Ref. 21.)

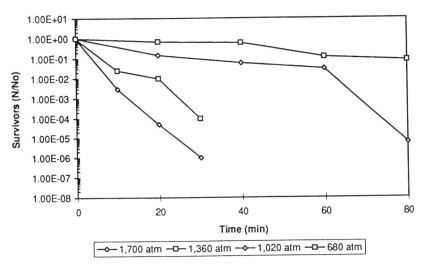

FIGURE 13 Inactivation of *V. parahaemolyticus* in 100 mM phosphate buffer with NaCl. (From Ref. 21.)

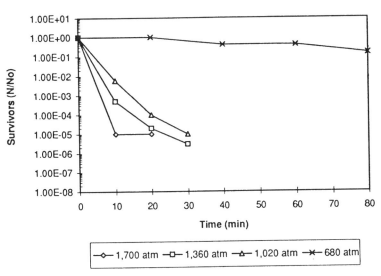

FIGURE 14 Inactivation of *V. parahaemolyticus* in commercially processed clam juice. (From Ref. 21.)

TABLE 5   Pressure Inactivation Times

Microorganism	Applied Pressure (atm)	Temperature (°C)	Time (min)	Viability
Bacteria in raw milk	2000	35	1800	One-log reduction
	5000	35	1800	Four-log reduction
	10,000	35	1800	Few cells survived
Bacillus anthracis	970	—	10	Vegetative cells and some spores killed
B. anthracis	2900	—	10	Some spores survived; their virulence was attenuated
Bacillus subtilis	5780–6800	—	5	Vegetative cells killed but not sterilized
Bacillus spores	1	93.6	60	Sterilization
	600	93.6	>240	Sterilization
Thermostable *B. subtilis* α-*amylase*	1	55	—	90% inactivation
Thermostable *B. subtilis* α-*amylase*	1000		1008	90% inactivation
C. diptheriae	4080–5440	—	5	Cells are sterilized
E. coli	2900	25–30	10	Most cells killed but not sterilized
E. coli	1		2160	9×10^8 stationary phase cells/mL reduces to 4×10^6 cells/mL
E. coli	1000	30	0	9×10^8 stationary phase cells/mL
			360	2×10^7 cells/mL
			720	3×10^5 cells/mL
			1080	6×10^3 cells/mL
			1440	1×10^2 cells/mL
			1800	4 cells/mL
			2160	Sterilization
E. coli	1000	40	720	Sterilization
E. coli	1000	20	720	Sterilization

TABLE 5   Continued

Microorganism	Applied Pressure (atm)	Temperature (°C)	Time (min)	Viability
Listeria monocytogenes	2380–3400	—	20	10^6 CFU/mL to less than 10 CFU/mL
Pseudomonas aeroginosa	1935	—	720	Cells are sterilized
Pseudomonas fluorescens	2040–3060	20–25	60	Cells are sterilized
Saccharomyces cerevisiae	5740	—	5	Cells are killed
S. albicans	3740–4080	—	5	Cells are killed
S. albicans	2040–2,80	—	60	Cells are killed
Salmonella typhimurium	4080–5440	—	5	Cells are sterilized
Salmonella typhimurium	2380–3400	—	30	3 log cycle reduction
Salmonella senftenberg	2380–3400	—	30	3–5-log cycle reduction
Serratia marcescens	5780–6800	20–25	5	Cells are sterilized
Staphylococcus aureus	2900	25–30	10	Most cells killed but not sterilized
Staphylococcus aureus antitoxin	1	65	30	85% denaturation
Staphylococcus aureus antitoxin	680	65	48	~85% denaturation
Streptococcus lactis	3400–4080	20–25	5	Cells are sterilized
Streptococcus spp.	1935	—	10	Sterilized
Vibrio cholera	1935	—	720	Cells are sterilized
Vibrio parahaemolyticus	1700	—	10-30	10^6 CFU/mL to less than 10 CFU/mL

pressures greater than 3000 atm. *E. coli* was inhibited at pressures greater than 4000 atm. More than 6 log CFU/g were inactivated at pressures greater than 4000 atm for 10 min. Fewer than 1 log CFU/g of *B. cereus* spores were inactivated by pressure treatment at 6000 atm. Pressure treatment causes leakage of substances possessing a UV absorption maximum at approximately 260 nm. Intensity of absorbance increased with an increase in pressure.

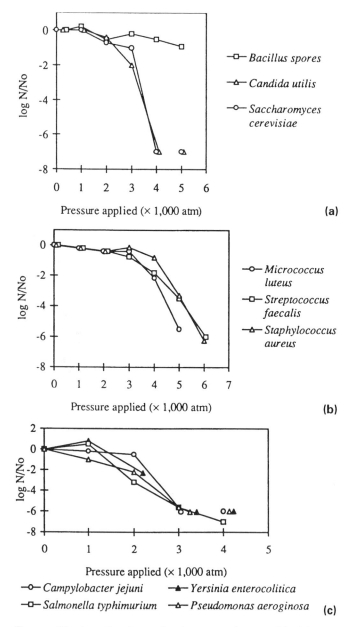

FIGURE 15   Inactivation of microorganisms with high pressure. (From Ref. 22.)

c. Inactivation of spores

One of the difficult operations in the preservation of food is the inactivation of bacterial spores. Though it is possible to inactivate spores by thermal treatment, it is not very desirable, because high temperature affects the quality of food. Johnson and Zobell (23) reported that an initial count of 8×10^4 /mL spores of *Bacillus subtilis* are inactivated when held at 93.6°C and 1 atm for 1 h. However, 4 h are required to inactivate the spores when the pressure is increased to 600 atm at 93.6°C. On the other hand, at low temperatures the rate of inactivation increases with an increase in pressure. At 25°C and 600 atm, the rate of inactivation is accelerated, with less than 10% of the initial count of spores being viable after 48 h.

Sale et al. (24) studied the inactivation of spores of *Bacillus* spp. at pressures ranging between 1000 and 8000 atm. The rate of inactivation is greater in the lower pressure region of 1000–3000 atm. Inactivation is enhanced when the temperature is increased to 70°C while the pressure is kept in the range of 1000 to 3000 atm.

Gould and Sale (25) reported that pressures of a few hundred atmospheres cause germination but are too low to inactivate the viable cells. Heat-activated spores are germinated more completely by pressure than are unheated spores. However, germination initiated at pressures greater than 1000 atm was influenced less by heat activation than germination initiated at lower pressures.

It was suggested that spores exist in a Donnan phase and that pressure triggers germination by influencing the equilibrium in the spore. Germination of spores is characterized by (26):

- An increase in heat and radiation sensitivity, respiratory activity, turbidity of the spore suspension, and stainability
- Loss of refractility
- Loss in weight
- Release of calcium and dipicolinic acid into the surrounding medium
- A decrease in molecular volume
- Structural and chemical changes

Under high pressure, the spores germinate to vegetative cells; the vegetative cells are then inactivated. At low temperatures, low pressure causes germination and heat sensitization but there is no appreciable inactivation of the germinated spores. Medium pressures cause considerable germination and a large proportion of the germinated spores are inactivated. High pressure causes less germination and only a small proportion of the germinated spores is inactivated. Above 65°C, inactivation of spores is caused by heat and not by pressure (24). The temperature and pressure ranges for germination and inactivation depend on the spore species. The loss of heat resistance of spores during initiation of germination is associated with increased hydration of cell, and gel to sol transformation (26).

The pressure inactivation of spores is strongly influenced by temperature and less strongly influenced by pH, water activity, and ionic strength. The optimum temperature for initiation of spore germination differs at different pressures. Inactivation of spores is greatest near neutrality and smallest at extreme pH levels. Pressure initiates the germination of spores optimally near neutral pH. Non-ionic solutes at low water activity (a_w) have little effect on inactivation of spores by pressure, whereas presence of ionic solutes NaCl and, more effectively, $CaCl_2$ decreases inactivation (24). Most spores cannot be germinated by high pressure in the absence of inorganic ions. The ions may affect the enzyme-catalyzed degradation of peptidoglycan during germination. The effective priority of ions on pressure-induced germination is H > K > Mn > Ca, Mg, Na (27). Ions also cause increased loss of heat resistance by spores in buffer solutions relative to distilled water suspensions at the same pressure and temperature (26).

Germination initiated by lower pressures is markedly affected by constituents of the suspending medium. For example, the amino acids alanine and riboside inosine are effective potentiators of pressure germination of *B. cereus* spores. The potentiators and inhibitors of germination become less effective in accelerating or preventing germination, respectively, as pressure increases. The inhibitors octyl alcohol (10 mM) and mercuric chloride (1 mM and 10 mM) completely prevent germination of *B. cereus* and *B. coagulans* at 400 and 600 atm, respectively. Sugar and salt are effective inhibitors of spore inactivation (5).

In the industry, thermoduric spores present in canned milk are inactivated by heating at 121°C for 20 min with addition of 300–500 SE's. However, this heat treatment causes quality decrease due to color changes and coagulation of casein. The thermoduric spore *Clostridium thermoaceticum* causes flat sour taste in canned liquid coffee during heat sealing. The spore is not inactivated when pressurized to 400-1000 atm for 20 min and 60 min at 20°C. However, combining pressure treatment with heating at 60°C for 60 min decreases the survivors to 100/mL (28). The combined effect of heat and high pressure on the inactivation of *B. stearothermophilus* is presented in Figure 16.

The method of pulse and oscillatory pressurization is more effective in spore inactivation than continuous pressurization. Six cycles (5 min/cycle) of oscillatory pressurization at 600 atm and 60°C decreased the population of *B. stearothermophilus* spores from 10^6 to 10^2 /mL. Strong inactivation ($10°$ count/mL) was achieved with six cycles (5 min/cycle) under 600 atm at 70°C. Spore inactivation can be greatly improved by increasing the oscillation cycle (28).

Pulse pressurization is useful in decreasing spore inactivation time. Only six cycles of pulse pressurization showed about one order decrease in survivor level. The total time exposed under 600 atm during the six-cycle pulse pressurization is less than 10 s. Spore inactivation decreased with increase in number of tubes in the pressurization chamber (28).

The mechanism of spore inactivation due to oscillatory pressurization is not clearly known. It is possible that the change in the properties of water (i.e., decrease in viscosity and surface tension at 70°C) may have an effect on destruction of spores by oscillatory pressurization. The surface of spores is completely ruptured after pressurization. Physical changes is the spore wall caused by an increase in temperature from 20° to 70°C may destroy spores. Also, a weakening of physical strength of spores may occur (28).

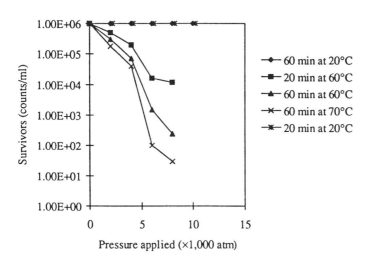

FIGURE 16 Inactivation of *B. stearothermophilus* spores by high pressure combined with or without heat. (From Ref. 28.)

d. Factors affecting microbial inactivation by high pressure

High pressure inactivation of microorganisms depends on pH, composition, osmotic pressure, and temperature of the media. At 1 atm, growth of *E. coli* is inhibited at pH 4.9 and 10.0; at 272 atm, inhibition occurs at pH 5.8 and 9.0. At 340 atm, *E. coli* growth is inhibited at pH 6.0 and 8.7. The threshold pressure tolerance is lower in nonnutrient salt solution than in the presence of essential amino acids and vitamins. High pressure alters the pH of the medium, which, in turn, affects the growth rate and inactivation rate. For example, at 0°C, seawater exhibits a pH of 8.10 at 1 atm. At 1100 atm, the pH of seawater decreases to 7.87. The extent of pressure effects on pH of solutions depends on the temperature and the electrolytes present. In the absence of Ca^{2+}, *B. subtilis* α-amylase was inactivated by pressure of approximately 500 atm. In the presence of 0.01 M Ca^{2+}, α-amylase was not appreciably inactivated at a pressure of 3000 atm. Bacteria are more sensitive to high pressure in mineral salt solutions and nutrient media (17). The sensitivity of barophobic bacteria to hypertonicity increases with an increase in pressure.

High pressure exhibits synergistic effects with moderate temperatures. Increased pressure retards inactivation of microorganisms at high temperatures. At 46.9°C, *E. coli* cells are inactivated less rapidly at 400 atm than at 1 atm (29). A heat-shock treatment of 51°C for 10 min protects *Saccharomyces cerevisiae* against damage at a subsequent high pressure treatment. Exposure of yeast cells to a pressure of 1500 atm for 60 min protects yeast against subsequent exposure to high temperature (30). Spores are inactivated within a few minutes or hours when pressurized to 600 to 1000 atm at 30°C, but survive for several months or years when held at the same pressure at low temperature (29).

The inactivation of microorganisms is dependent on water activity. Pressure inactivation of *Rhodotorula rubra* is hindered when water activity is less than 0.94. At a_w 0.96, a 7-log cycle/mL reduction in the microbial count is obtained, but at a_w 0.91 no reduction is obtained (Fig. 17). Soluble solids prevent inactivation of *S. cerevisiae*, *Aspergillus awamori*, *S. bayanus*, *Pichia membranaefaciens*, and *Mucor plumbens* (32).

e. Kinetics of microbial inactivation

Carlez et al. (33) reported that high pressure inactivation of *Citrobacter freundii*, *Pseudomonas fluorescens*, and *Listeria innocua* inoculated in minced beef muscle follows first-order kinetics. The inactivation kinetics for *C. freundii* can be described by (33):

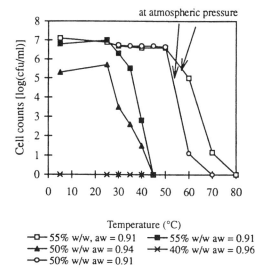

FIGURE 17 Sucrose concentration (% w/w), water activity (a_w), temperature, and pressurization of *Rhodotorula rubra* at 4000 atm for 15 min. (From Ref. 31.)

$$\text{Log } n = 7.017 - 0.068\, t \tag{1}$$

where n is the number of survivors after pressurization and t is the time of pressurization at 230 atm. The time required to reduce the survivor population by one log cycle at 150 atm and 20°C, $D_{150\,\text{atm},\,20°C}$, is 14.7 min. At 150 atm and 20°C, *P. fluorescens* inactivation can be described by:

$$\text{Log } n = 6.754 - 0.042\, t \tag{2}$$

$D_{150\,\text{atm},\,20°C}$ for *P. fluorescens* is 23.8 min. At 330 atm and 20°C, the inactivation of *L. innocua* can be described by:

$$\text{Log } n = 7.171 - 0.155\, t \tag{3}$$

and $D_{330\,\text{atm},\,20°C}$ for *L. innocua* is 6.5 min.

A similar apparent first-order inactivation of *E. coli* was observed in physiological saline solution at 20°C and pressures of 200, 250, 300, and 350 atm. First-order inactivation of *E. coli* was also observed in aqueous solution at 40 and 50°C and 250 atm. However, inactivation at 25°C and 250 atm deviated slightly from the linear equation describing first-order kinetics because a small population of the bacteria was not inactivated even after long periods of pressurization. A similar deviation from first-order kinetics was also observed after pressurization of coliforms in bovine colostrum at 20°C and 200 atm for 4 h. The bacteria surviving at the end of 4 h were inactivated at a pressure larger than 200 atm (33).

The initiation of germination and inactivation of phosphate-buffered *Bacillus pumilus* spores followed first-order kinetics at 25°C (26).

B. High Pressure and Enzymatic Reactions

Exposure to high pressure results in the activation or inactivation of enzymes. The activity of succinate, formate, and malate dehydrogenases in *E. coli* decreases with an increase in pressure. The dehydrogenases are completely inactivated when subjected to a pressure of 1000 atm for 15 min at 27°C (34). Inactivation of enzymes is influenced by the pH, substrate concentration, subunit structure of enzyme, and temperature of pressurization (35).

The activity of *E. coli* aspartase increases when pressure is raised to 680 atm. Aspartase activity ceases when pressure is increased to 1000 atm (34). However, the thermal denaturation of aspartase at temperatures in the range of 46–56°C can be prevented by increasing the pressure to 1000 atm. Reactions catalyzed by thermolysin or cellulase are stimulated by high pressure (36). The activity of thermolysin increases by about 15 times when pressurized to 1000 atm (37). Trypsin and carboxypeptidase Y can be inactivated by high pressure (38).

Tetrameric lactate dehydrogenase from *B. stearothermophilus* exhibits unusual stability toward high pressure. Partial dissociation of the dimer and reassociation after pressure release occurs. The reassociated enzyme is indistinguishable from the physicochemical and enzymological properties of the native enzyme (39).

Pressurization at 1,000 and 2,000 atm causes hardly any inactivation of pectinestearase. Pectinestearase in juices such as Satsuma mandarin juice is inactivated when pressurized to 3000–4000 atm. Purified pectinestearase is also inactivated at 3000 atm or higher. The inactivation is irreversible, and the pectinestearase is not reactivated during storage at 0°C or transportation (Fig. 18). Soluble solids such as sugars, proteins, and lipids exert a protective action against pectinestearase inactivation by high pressure or by heat (32). Enzyme inactivation takes place as a result of pressure alteration of intramol-

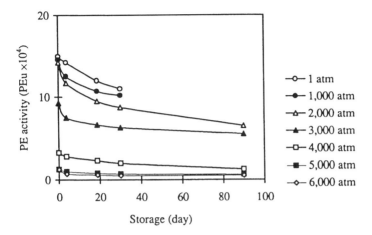

FIGURE 18 Decrease in pectinestearase (PE) activity of pressurized Satsuma mandarin juice during storage at 0°C. (From Ref. 32.)

ecular structures or conformational changes at the active site. Inactivation of some enzymes pressurized to 1000–3000 atm is reversible. Reactivation after decompression depends on the degree of distortion of the molecule. The chances of reactivation decrease with an increase in pressure beyond 3000 atm (40,41).

The proteolytic activity of enzymes in meat is enhanced by the application of high pressures (42). The total activities of cathepsin B, D, L and acid phosphatase in the muscle increase when subjected to pressures ranging between 1000 and 5000 atm for 5 min at 2°C. Cathepsin H and aminopeptidase B are resistant to pressure treatment (Fig. 19). An increase in activity of cathepsin B1 may account, in part, for the tenderization of meat by pressure-heat treatment (43).

High pressure can exert a drastic change in the reaction rate or specificity of an enzyme and can be used to selectively digest one protein in a set of proteins during proteolysis (37).

The activity of polyphenol oxidase increases five times when slices of Bartlett pears are pressurized at 4000 atm and 25°C for 10 min (Fig. 20). Further increase in pressure does not increase the enzyme activity. On the other hand, pressurization of homogenates of apples, bananas, or sweet potatoes did not result in activation of polyphenol oxidase (44).

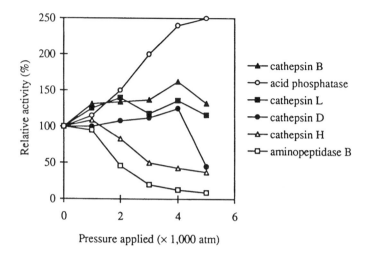

FIGURE 19   Pressure and proteolytic enzyme activity in meat. (From Ref. 42.)

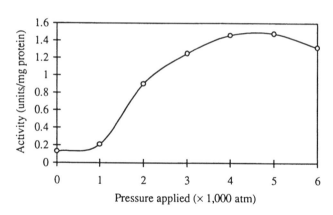

FIGURE 20   Activation of polyphenol oxidase (PPO) from Bartlett pears by high pressure. (From Ref. 44.)

C. High Pressure and Biochemical Reactions

The biochemical reactions most affected by high pressure are the reactions with reactants that undergo either a decrease or an increase in the volume (45). Pressure causes a decrease in the available molecular space, or an increase in chain interactions. Reactions involved with formation of hydrogen bonds are favored by high pressure because bonding results in a decrease in volume (35). However, Masson (46) reports that hydrogen bonds are insensitive to pressure.

High pressure denatures protein molecules. Pressure denaturation of proteins is a complex phenomenon depending on the protein structure, pressure range, temperature, pH, and solvent composition. Oligomeric proteins are dissociated by relatively low pressures (2000 atm), whereas single chain protein denaturation occurs at pressures greater than 3000 atm. Pressure-induced denaturation is sometimes reversible, but renaturation after pressure release may take a long time. The effects of pressure on protein stability are governed by Le Chatelier's principle: positive/negative changes in volume with an increase in pressure cause an equilibrium shift toward bond breakage/bond formation. The main targets of pressure are the electrostatic and hydrophobic bonds in protein molecules. High pressure causes deprotonation of charged groups, disruption of salt bridges, and hydrophobic bonds, thereby resulting in conformational and structural changes in proteins. Structural transitions are accompanied by large hydration changes. Hydration changes are the major source of volume decreases associated with dissociation and unfolding of proteins (46).

Pressure-induced denaturation of proteins is different from heat-induced denaturation. High pressure disrupts hydrophobic and ion-pair bonds of the protein molecule. The unfolding of protein molecules reduces volume of the protein by 2% (47). Protein unfolding persists for at least 8 days in milk stored at 5°C. Unfolding of protein results in alterations of the functional properties (48). Heat-induced denaturation is caused by the formation or destruction of covalent bonds (16).

Fermentation reactions are retarded at high pressures. The decomposition products obtained by high pressure fermentation are quite different from products obtained at 1 atm. Raw milk does not sour in 12 days when kept at 700 atm (16). Pressure of 13,710 atm applied for 1 h postpones souring of raw milk for about 4 days. Similar results were obtained by application of 4,567 atm for 10–12 h, and further improvement was obtained by combining pressure treatment with temperature (48).

The high acidity in yogurt caused by continuous fermentation during storage can be prevented by pressure treatment of yogurt at 2000–3000 atm for 10 min at 10°C. The lactic acid bacteria are maintained at the initial population and further growth is prevented (36).

The formation of milk curd by rennet is accelerated when milk is pressurized to less than 400 atm. The primary phase in milk curd formation (i.e., casein degradation) is not affected by high pressure. The secondary phase, core formation, for initiating milk curdling is retarded by high pressure; the tertiary phase—milk curdling—is enhanced by high pressure. Milk curd formation is suppressed when pressure is increased to 600–1300 atm. After 70 min of compression at 1300 atm and 35°C, the extent of proteolysis increases up to about 2%, and the change in proteolytic activity of rennet was negligible. One of the reasons for delay of secondary phase development may be suppression of collisions between casein micelles destabilized by the partial degradation of κ-casein by rennet. At pressures greater than 600 atm, the secondary phase was a rate-limiting process that delayed the initiation of tertiary phase, resulting in a suppression of milk curdling (49).

D. High Pressure and Microbial Cells

Nucleic acids are more baroresistant than protein molecules. The structure of DNA is largely a result of hydrogen bonding. Because high pressure favors hydrogen bonding, the DNA molecules are more stable at high pressure than proteins, whereas high temperature causes denaturation of DNA molecules. Hedén (45) observed no denaturation of DNA when *Bacillus subtilis* DNA solutions (0.002–0.04%, pH 4.8–9.9) were subjected to pressures up to 10,000 atm at room temperature. However, DNA transcription and replication are disrupted by high pressure due to the involvement of enzymes. The disruption is reversible at low pressure and irreversible at high pressures (50).

The cell membrane is influenced by changes in the external environment. The most important functions of cell membranes are: (a) to provide barriers to diffusion, especially of charged solutes, (b) to support receptors, enzymes, and ion channels; the ion channels are involved in selectively transporting solutes across the membrane, (c) to provide lipid precursors for the cells' internal signaling system, and (d) to provide shape to the cell (51).

High pressure denatures the proteins and reduces the size of phospholipids in the cell membrane. Denaturation of proteins inhibits the uptake of amino acids essential for cell growth. High pressure increases the permeability of the cell membrane and the contents of the cell leak out, disrupting the functioning of the cell. If the applied pressure is relatively low, the cell regains the original permeability. The destruction of the cell wall is irreversible when the applied pressure is relatively high and results in cell inactivation (16).

IV. APPLICATIONS OF HIGH PRESSURE IN FOOD PROCESSING

High pressure technology can be applied to extend the shelf life of foods and modify the texture and sensory properties of foods. Tables 6 and 7 illustrate some of the possible applications of high pressure processing in the food industry (6).

TABLE 6 Applications of High Pressure Technology to Food Processing

Effect	Solid Food					Liquid Food	
	Fish	Meat	Eggs	Rice, starches	Soy bean protein	Milk	Natural Juice
Prolongation of storage time						●[a]	●
Prevention of microbial contamination	◆[b]	◆	◆	◆	◆	◆	●
Development of new foodstuffs	●	●	●	●	●		
Manufacture of partially cooked food	●	●	●	●			

●[a] Applicability is large.
◆[b] Application exists.
Source: From Ref. 6.

TABLE 7 Applications of High Pressure Technology to Processed Foods

Effect	Desserts	Pickles	Cheese	Seasonings	Spices
Prolongation of storage time	●[a]	●	●	●	●
Prevention of microbial contamination	◆[b]	◆	◆	◆	◆

●[a] Applicability is large.
◆[b] Application exists.
Source: From Ref. 6.

High pressure processing can be used for increasing the shelf life of foods, for thawing frozen foods, and for storage of foods without freezing. The shelf life of foods is increased by the inactivation of microorganisms, spores, and undesirable enzymes using the appropriate pressures. It is desirable to thaw frozen foods uniformly. High isostatic pressure can thaw foods uniformly. Pressure thawing is faster when foods contain more soluble solids such as NaCl or sugars (52).

A. Sensory Quality of Foods

Exposing foods to high pressure results in sensory changes dependent on the type of food. For example, the internal structure of tomatoes becomes tough, tissues of chicken and fish fillets become opaque, and pre-rigor beef is tenderized (35). Because fresh meat is not tender, it must be stored at refrigerated temperature for about 2 weeks before it becomes tender, but with high pressure processing, tenderization of fresh meat requires only 10 min (53).

Pressure-treated bovine longissimus muscle has a shorter sarcomere length and lower pH and Warner-Bratzler values. Physical changes in muscle tissue induced by high pressure include separation of sarcolemmal and endomysial sheaths, contraction bands, and disruption of myofibrillar and intermyofibrillar spaces. Disappearance of glycogen granules, appearance of swollen mitochondria and sarcoplasmic reticulum, and in some cases rupture of mitochondria occurs on pressurization. The interaction between chemical and physical effects on muscle tissue accounts for the tenderizing effect of pressure treatment (54). Also, the increase in activity of the endogenous muscle proteolytic enzymes due to high pressure may contribute to the tenderization of meat (42).

The alteration of the structure of starch and protein by high pressure can be utilized so rice can be cooked in few minutes (35). Grapefruit juice manufactured by high pressure technology does not possess the bitter taste of limonene present in conventional thermal processing (53). Peaches and pears processed at 4100 atm for 30 min remain commercially sterile for 5 years (35). Pressure treatment of non-pasteurized citrus juices provide a fresh-like flavor with no loss of vitamin C and a shelf life of approximately 17 months (16).

In Japan, high pressure processing is utilized for the manufacture of jams and sauces from strawberry fruit and orange marmalade. The desired plastic container is filled with a mixture of the raw materials consisting of fruits, fruit juice, sugar, and acidulants. The container is sealed and subjected to a pressure of 4000 to 6000 atm for 1–30 min. Strawberry jam can be obtained by pressurization at 4000 atm for 15 min, and strawberry puree obtained by pressurization at 4000 atm for 10 min. The jams obtained by high pressure processing retain the taste and color of fresh fruit, unlike conventional jams produced by cooking. Pressurization allows the permeation of sugar solution into the fruits as well as commercial preservation of the jam (55).

During the pressure treatment, fruits such as pears and persimmons become soft, transparent, and sweet. However, the color of fruits and vegetables, including pears, apples, potatoes, and sweet potatoes rapidly darkens after high pressure treatment. In pear slices, the color darkens within 30 min after pressurizing at 4000 atm and 25°C for 10 min. The activity of polyphenol oxidase (PPO) in pressure-treated fruits and vegetables is five times greater than the activity of PPO in fresh fruits and vegetables. The PPO activity does not increase with repeated pressurization (44).

The texture of potatoes, root vegetables, and sweet potatoes becomes pliable when subjected to a pressure of 5000 atm at room temperature for 15 min. However, enzymatic browning is accelerated in potatoes by pressurization (56).

Pressure inactivation of PPO depends on the medium, fruit, or vegetable being treated. For example, PPO is not inactivated in deionized water, water containing Ca^{2+}, or water containing carbon dioxide at a pressure of 4000 atm and a temperature of 50°C. PPO is inactivated in a solution of 0.5% citric acid (30). Complete inactivation of PPO is achieved at 4000 atm and 20°C within 15 min in a 0.5% citric acid solution.

High pressure may increase the benzaldelyde content of fruits. Increased benzaldehyde concentration may contribute to the flavor quality of fruits (57).

B. Gelation of Proteins

Egg yolk subjected to a pressure of 4000 atm for 30 min at 25°C forms a gel. A pressure of 5000 atm renders egg white partially coagulated and opaque; a pressure of 6000 atm causes complete gelation. Pressure-induced gels of egg white possess a natural flavor, display no destruction of vitamins and amino acids, and are more easily digested when compared with heat-induced gels. The gels retain the original color of the yolk or the white and are soft, lustrous, and adhesive when compared with heat-induced gels. While the strength of the gels increase, the adhesiveness decreases with an increase in the applied pressure. However, the hardest gel formed by high pressure (5000 atm) treatment exhibits one sixth the strength of heat-induced gels. Gumminess of pressure-induced gels is considerably less than gumminess of heat-induced gels. Gels of egg white produced at 6000 and 7000 atm deform readily without fracture. Cohesiveness of pressure-induced gels increase with increases in applied pressure. The force deformation curves of pressure- and heat-induced gels of egg yolk and egg white are presented in Figures 21 and 22, respectively (58).

Boiled eggs often exhibit a sulfur flavor and contain lysinoalanine produced during cooking. Pressure-treated eggs do not exhibit the sulfur flavor or contain lysinoalanine. Lysinoalanine forms a three-dimensional network in the intestine and inhibits the activation of proteolytic enzymes; therefore, the availability of amino acids to the human body is reduced. Processing with high pressure does not affect the riboflavin, folic acid, or thiamine in eggs, whereas boiling breaks down these vitamins (15).

The melting point of carrageenan, ovalbumin, and soy protein gels decreases linearly with increasing pressure, indicating formation of less stable gels under high pressure. The melting point of agarose and gelatin gels increases with an increase in pressure (59).

In Japan, a hydrostatic pressure of 4000 atm is used to induce gelation of pollack-, sardine-, skipjack-, and tuna-based surimi. Squid-based surimi is obtained by pressurization of extracted muscle protein to 6000 atm. Pressure-induced surimi gels are organoleptically superior to heat-induced surimi gels (16). Gelation can be used for adhesion-binding of small size muscles or fish fillets, restructuring of minced fish or deboned meat, and molding of surimi or

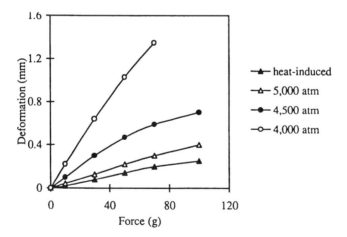

FIGURE 21   Force deformation curves of pressure (4000 to 5000 atm) and heat-induced gels of egg yolk. (From Ref. 58.)

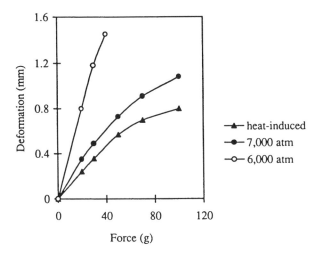

FIGURE 22 Force deformation curves of pressure (6000 and 7000 atm) and heat-induced gels of egg yolk. (From Ref. 58.)

pieces of gelled surimi into seafood analogues. The possibility of obtaining acceptable gels simultaneous with commercial sterilization at a temperature as low as 0°C is of tremendous practical interest to the surimi industry (5).

The mechanism of high pressure–induced gelation is different from heat-induced gelation. Gelation by high pressure treatment is attributed to a decrease in the volume of the protein solution. On the other hand, the application of heat results in violent movement of protein molecules leading to the destruction of noncovalent bonds, denaturation, and formation of a random network. The rearrangement of water molecules around amino acid residues in pressure-induced gels produces glossy and transparent gels compared to opaque gels obtained by high temperatures (58).

C. Quality and Functionality Improvement

Under the influence of high pressure, covalent bonds remain intact, whereas noncovalent bonds break down. Pressure treatment increases the digestibility of meat proteins. The apparent biological value of protein and the protein efficiency ratio of pressure-treated meat are equivalent to meat proteins tenderized at atmospheric pressure (60).

Enzymes are generally inactivated in vegetables by hot water blanching. Disadvantages of blanching include thermal damage, leaching of nutrients, and possible environmental pollution due to the production of high biochemical oxygen demand effluent. Water blanching reduces total counts of microorganisms by three log cycles, whereas ultrahigh pressure (UHP) reduction in total counts was generally four log cycles. During hot water potato blanching, a continuous increase of potassium leaching occurs. The greatest potassium loss was observed when water was the immersion medium. Leaching of potassium ions decreased with increasing temperature in water, but it increased with temperature in a $CaCl_2$ solution. UHP treatment can fulfill the requirements of hot water blanching while avoiding the mineral leaching and the accumulation of waste water. High pressure treatment is potentially a process producing less effluent because less water is required than in hot water blanching. High pressure treatment may also process vacuum-packaged foods (56).

High pressure can be combined with the inhibitory effect of carbon dioxide in the sterilization of fresh and freeze-dried herbs and spices. Microorganisms are not inactivated if the spices are dry. The presence of moisture is important to the antimicrobial action of carbon dioxide. Exposure of spices to carbon dioxide at 55 atm and 45°C for 2 h inactivates microorganisms in fresh chives, thyme, parsley, and mint. Thyme, mint, and chives exhibit enhanced aromas and flavor after pressure treatment (61).

High pressure processing thaws frozen foods much faster than conventional thawing methods. Two kilograms of frozen beef thaws in 80 min when pressurized to 2000 atm; thawing at atmospheric pressure takes 7 h. The flavor and juiciness of the pressure-thawed beef are equivalent to the flavor and juiciness obtained by thawing under low humidity at 5°C. The surface of the pressure-thawed beef is slightly discolored. Meat stored at 3°C after pressurization for 20 min at 2000 atm and 20°C begins to spoil after 2 days. The spoilage of meat begins after 6 days when it is pressurized at 3000 atm and after 6 to 13 days when pressurized at 4000–4,500 atm (62). However, meat remains unaffected by microbial spoilage under refrigerated temperatures after a pressure treatment for 1 h at 5400 atm and 52°C.

Microflora initial population (10^6 cells/mL) of milk is reduced by 99.95% at 8200 atm and 13°C for treatment times from 40 s to 83 min (63).

Gelatinization is the transition of starch granules from the birefringent crystalline state to a non-birefringent, swollen state. Starch can be gelatinized using pressure or heat. The pressure at which starch gelatinizes depends on the source of starch. Gelatinization may be stimulated by increased temperatures of pressurization (64). High pressure may also produce an upward shift of gelatinization temperature of about 3–5° per 1000 atm. Pressure higher than 1500 atm does not further enhance gelatinization temperature. The effect of high pressure on gelatinization is the result of the stabilization of hydrogen bonds that maintain the starch granule in the original state (65).

The characteristic changes of pressure-gelatinized starches during cold storage are slower than changes of heat-gelatinized starches during cold storage. Heat treatment breaks down the natural structure of starch during gelatinization. Storage at low temperatures partially restores the natural structure of starch. Pressurization causes swelling of starch granules while still maintaining the granular structure (66).

Exposure to high pressures unfolds protein molecules. Unfolding results in alterations of the functional properties of the protein (48). Foaming, emulsifying, gelling, and water-binding capacity of proteins may be influenced. Proteins treated with high pressures may lead to the development of a range of functional food ingredients prepared from food proteins by controlled unfolding (48). The improvement of emulsifying properties of pressure-treated proteins is pH dependent. A solution of ovalbumin exhibits high emulsion stability and emulsifying activity at pH 12.5 under a pressure of 6000 atm and temperature of 25°C. The emulsifying activity decreases when pH is decreased to 5.5 near the isoelectric point of ovalbumin. However, the emulsifying properties of casein and bovine serum albumin are not significantly improved by pressure treatment. The difference in the behavior of the proteins may be attributed to differences in the S-S bridges in the proteins (67).

D. Miscellaneous

High pressure can be used to improve the extraction yield of important metabolites such as flavors or pigments at ambient temperature. For example, amaranthin pigment can be extracted from the cells of *Chenopodium rubrum* by pressurization. Pressure treatment of the cells at 2500 atm for 10 min causes 99% of the pigment to be released into the culture medium (31).

High pressure retards pigment formation in bacteria. For example, pigment formation is reversibly retarded in *Serratia marcescens* and *Staphylococcus aureus* at 2900 atm and 30°C. The normally red *Serratia marinorubra* commonly produces orange, yellow, cream, and white color variants in decreasing order at atmospheric pressure. At 300 atm, the order of color variants is reversed, with more white or colorless and fewer red color variants (17).

High pressure processing can be used to improve the selective proteolysis of proteins. For example, β-lactoglobulin is preferentially digested by thermolysin after a pressure treatment of 2000 atm, whereas α-lactalbumin remains intact. Selective elimination of β-lactoglobulin from bovine milk whey while maintaining α-lactalbumin is important when whey concentrates are used as an additive in modified milk for infants (68).

The presence of ice nucleation–active bacteria enhance frost injury to crops at low temperatures. Many investigations are directed toward finding methods to inhibit the activity of these bacteria. Cells of ice nucleation–active bacteria are inactivated when pressures greater than 3000 atm are applied (69).

The melting point of triglycerides increases as the pressure increases. Lipids present in the liquid form at room temperature and pressure crystallize under high pressures. High pressure enhances the formation of dense and stable fat crystals; β crystals are formed in preference to β′, and β′ crystals are formed in preference to α crystals (52).

Acid hydrolysis of proteins is enhanced under high pressure treatment, whereas hydrolysis of corn starch and locust bean gum is not affected. Maillard reactions involving xylose lysine, xylose β-alanine, or glutaraldehyde β-alanine are inhibited by high pressure treatment (52).

High pressure treatment at 1000–4000 atm and 4–20°C for 30 min inactivates lipase in sardine meat. Inactivation of lipase prevents accumulation of free fatty acids, which contribute to off-flavors. Pigments such as carotenoids, chlorophyll, and anthocyanins resist changes during pressure treatment. However, myoglobin is pressure sensitive, and therefore meat treated with high pressures loses the bright red color (52).

V. FINAL REMARKS

High pressure technology is a very promising nonthermal food preservation method. High pressure can be used not only to preserve foods, but also to improve the rheological and functional properties of foods. An important aspect of high pressure technology is the inactivation of enzymes while nutrients and flavors are retained in the food, giving a fresh-like flavor and texture to pressure-processed foods. The technical difficulty of fabricating pressure vessels that will tolerate very high pressures has limited the commercialization of high pressure technology. However, efforts are being made to overcome this difficulty.

REFERENCES

1. B. Mertens and D. Knorr, Development of non thermal processes for food preservation. *Food Technol.* 46 (5): 124-133 (1992).
2. F. Zimmerman and C. Bergman, Isostatic pressure equipment for food preservation. *Food Technol.* 47 (6): 162-163 (1993).

3. G. Deplace and B. Mertens, The commercial application of high pressure technology in the food processing industry, *High Pressure and Biotechnology* (R. Hayashi, K. Heremans, and P. Masson, eds.), Colloque INSERM John Libbey Eurotext Ltd., 1992, Vol. 224, p. 469.
4. B. Mertens and G. Deplace, Engineering aspects of high pressure technology in the food industry. *Food Technol.* 47(6): 164-169 (1993).
5. J. C. Cheftel, Applications des hautes pressions en technologie alimentaire. *Ind. Alim. Agric.* 108: 141-153 (1991).
6. Anonymous, Mitsubishi Heavy Industries, Technical Data (1992).
7. K. Hori, Y. Manabe, M. Kaneko, T. Sekimoto, Y. Sugimoto, and T. Yamane, The development of high pressure processor for food industries, *High Pressure and Biotechnology* (R. Hayashi, K. Heremans, and P. Masson, eds.), Colloque IN-SERM John Libbey Eurotext Ltd., 1992, Vol. 224, p. 499.
8. T. Kanda, T. Yamauchi, T. Naoi, and Y. Inoue, Present status and future prospects of high pressure food processing equipment, *High Pressure and Biotechnology* (R. Hayashi, K. Heremans, and P. Masson, eds.), Colloque INSERM John Libbey Eurotext Ltd., 1992, Vol. 224, p. 521.
9. Anonymous, ABB Autoclave Systems, Technical Data (1993).
10. R. J. Swientek, High hydrostatic pressure for food preservation. Technologies for tomorrow. *Food Processing*, Nov. (1992).
11. Anonymous, ABB Autoclave Systems, Technical Data (1994a).
12. Anonymous, Engineered Pressure Systems, Technical Data (1994b).
13. C. E. Morris, High pressure builds up. *Food Eng.* Oct.: 113-120 (1993).
14. M. Masuda, Y. Saito, T. Iwanami, and Y. Hirai, Effects of hydrostatic pressure on packaging materials for food, *High Pressure and Biotechnology* (R. Hayashi, K. Heremans, and P. Masson, eds.), Colloque INSERM John Libbey Eurotext Ltd., 1992, Vol. 224, p. 545.
15. R. Hayashi, Application of high pressure to food processing and preservation: philosophy and development, *Engineering and Food* (W. E. L. Spiess and H. Schubert, eds.), Elsevier Appl. Sci., London, 1989, Vol. 2, p. 815.
16. D. Farr, High pressure technology in the food industry. *Trends Food Sci. Technol.* 1: 14-16 (1990).
17. C. E. Zobell, Pressure effects on morphology and life processes of bacteria, *High pressure effects on cellular processes* (A. M. Zimmerman, ed.), Acad. Press., New York and London (1970).
18. C. E. Zobell and A. B. Cobet, Growth, reproduction, and death rates of *Escherichia coli* at increased hydrostatic pressures. *J. Bacteriol.* 84: 1228-1236 (1962).
19. J. A. Kitching, Effects of high hydrostatic pressure on the activity of flagellates and ciliates. *J. Experimental Biol.* 34: 494-510 (1957).
20. C. Metrick, D. G. Hoover, and D. F. Farkas, Effects of high hydrostatic pressure on heat-resistant and heat-sensitive strains of Salmonella. *J. Food Sci.* 54(6): 1547-1564 (1989).

21. M. F. Styles, D. G. Hoover, and D. F. Farkas, Response of *Listeria monocytogenes* and *Vibrio parahaemolyticus* to high hydrostatic pressure. *J. Food Sci.* 56 (5): 1404-1407 (1991).

22. T. Shigehisa, T. Ohmori, A. Saito, S. Taji, and R. Hayashi, Effects of high hydrostatic pressure on characteristics of pork slurries and inactivation of microorganisms associated with meat and meat products. *Int. J. Food Microbiol.* 12: 207-216 (1991).

23. F. H. Johnson and C. E. Zobell, The retardation of thermal disinfection of *Bacillus subtilis* spores by hydrostatic pressure. *J. Bacteriol.* 57: 353-358 (1949).

24. A. J. H. Sale, G. W. Gould, and W. A. Hamilton, Inactivation of bacterial spores by hydrostatic pressure. *J. Gen. Microbiol.* 60 (3): 323-334 (1969).

25. G. W. Gould and A. J. H. Sale, Initiation of germination of bacterial spores by hydrostatic pressure. *J. Gen. Microbiol.* 60: 335-346 (1970).

26. J. G. Clouston and P. A. Wills, Kinetics of initiation and germination of *Bacillus pumilus* spores by hydrostatic pressure. *J. Bacteriol.* July: 140-143 (1970).

27. G. R. Bender and R. E. Marquis, Sensitivity of various salt forms of *Bacillus megaterium* spores to the germinating action of hydrostatic pressure. *Can. J. Microbiol.* 28: 643-649 (1982).

28. I. Hayakawa, T. Kanno, M. Tomita, and Y. Fujio, Application of high pressure for spore inactivation and protein denaturation. *J. Food Sci.* 59(1): 159-163 (1994).

29. J. Kim, Disinfection by increased hydrostatic pressure. *Dev. Ind. Microbiol.* 24: 519-525 (1983).

30. H. Iwahasi, S. C. Kaul, K. Obuchi, and Y. Komatsu, Induction of barotolerance by heat shock treatment in yeast. *FEMS Microbiol. Lett.* 80: 325-328 (1991).

31. D. Knorr, Effects of high hydrostatic pressure on food safety and quality. *Food Technol.* 47(6): 156-161 (1993).

32. H. Ogawa, K. Fukuhisa, Y. Kubo, H. Fukumoto, Inactivation effect of pressure does not depend on the pH of the juice. *Agric. Biol. Chem.* 54 (5): 1219-1225 (1990).

33. A. Carlez, J. P. Rosec, N. Richard, and J. C. Cheftel, High pressure inactivation of *Citrobacter freundii*, *Pseudomonas fluorescens* and *Listeria innocua* in inoculated minced beef muscle. *Lebensm. Wiss. U. Technol.* 26: 357-363 (1993).

34. R. Y. Morita, Effect of hydrostatic pressure on succinic, malic and formic dehydrogenases in *Escherichia coli. J. Bacteriol.* 74: 251-255 (1957).

35. D. G. Hoover, C. Metrick, A. M. Papineau, D. F. Farkas, and D. Knorr, Biological effects of high hydrostatic pressure on food microorganisms. *Food Technol.* 43(3): 99-107 (1989).

36. D. G. Hoover, Pressure effects on biological systems. *Food Technol.* 47(6): 150-155 (1993).

37. S. Kunugi, Effect of pressure on activity and specificity of some hydrolytic enzymes, *High pressure and Biotechnology* (R. Hayashi, K. Heremans, and P. Masson, eds.), Colloque INSERM John Libbey Eurotext Ltd., 1992, Vol. 224, p. 129.

38. S. Kunugi, M. Fukuda, and N. Ise, Pressure dependence of trypsin-catalyzed hydrolyses of specific substrates. *Biochim. Biophys. Acta* 704: 107-113 (1982).
39. K. Müller, T. Seifert, and R. Jaenicke, High pressure dissociation of lactate dehydrogenase from *Bacillus stearothermophilus* and reconstitution of the enzyme after denaturation in 6 M guanidine hydrochloride. *Eur. Biophys. J.* 11: 87-94 (1984).
40. R. Jaenicke, Enzymes under extreme conditions. *Ann. Rev. Biophys. Bioeng.* 10: 1 (1981).
41. C. Suzuki and K. Suzuki, The gelation of ovalbumin solutions by high pressure. *Arch. Biochem. Biophys.* 102 (3): 367 (1963).
42. N. Homma, Y. Ikeuchi, and A. Suzuki, Effect of high pressure treatment on the proteolytic enzymes in meat. *Meat Sci.* 38: 219-228 (1994).
43. L. B. Kurth, Effect of pressure-heat treatments on cathepsin B1 activity. *J. Food Sci.* 51(3): 663-667 (1986).
44. M. Asaka and R. Hayashi, Activation of polyphenol oxidase in pear fruits by high pressure treatment. *Agric. Biol. Chem.* 55(9): 2439-2440 (1991).
45. C. G. Hedén, Effects of high hydrostatic pressure on microbial systems. *Bacteriol. Rev.* 28: 14-29 (1964).
46. P. Masson, Pressure denaturation of proteins, *High Pressure and Biotechnology* (R. Hayashi, K. Heremans, and P. Masson, eds.), Colloque INSERM John Libbey Eurotext Ltd., 1992, Vol. 224, p. 89.
47. A. A. Zamyatnin, Protein volume in solution. *Prog. Biophys. Molec. Biol.* 24 (1): 107 (1972).
48. D. E. Johnston, B. A. Austin, and R. J. Murphy, Effects of high hydrostatic pressure on milk. *Milchwissenchaft* 47(12): 760-763 (1992).
49. K. Ohmiya, K. Fukami, S. Shimizu, and K. Gekko, Milk curdling by rennet under high pressure. *J. Food Sci.* 52 (1): 84-87 (1987).
50. J. V. Landau, Induction, transcription and translation in *Escherichia coli*: a hydrostatic pressure study. *Biochim. Biophys. Acta* 149: 506-512 (1967).
51. A. G. Macdonald, Effects of high pressure on natural and artificial membranes, *High pressure and Biotechnology* (R. Hayashi, K. Heremans, and P. Masson, eds.), Colloque INSERM John Libbey Eurotext Ltd., 1992, Vol. 224, p. 67.
52. J. C. Cheftel, Effects of high hydrostatic pressure on food macromolecules: an overview, *High Pressure and Biotechnology* (R. Hayashi, K. Heremans, and P. Masson, eds.), Colloque INSERM John Libbey Eurotext Ltd., 1992, Vol. 224, p. 195.
53. S. Nagatsuji, The fat of the land under pressure. *Look Japan*, Oct.: 28-29 (1992).
54. E. A. Elgasim and W. H. Kennick, Effect of hydrostatic pressure on meat microstructure. *Food Microstructure* 1: 75-82 (1982).
55. Y. N. Horie, K. I. Kimura, and M. S. Ida, Jams treated at high pressure. U. S. Patent 5,075,124 (1991).
56. M. N. Eshtiaghi and D. Knorr, Potato cubes response to water blanching and high hydrostatic pressure. *J. Food Sci.* 58(6): 1371-1374 (1993).

57. H. Sumitani, S. Suekane, A. Nakatani and K. Tatsuka, Changes in composition of volatile compounds in high pressure treated peach. *J. Agric. Food Chem.* 42: 785-790 (1994).

58. M. Okamoto, Y. Kawamura, and R. Hayashi, Application of high pressure to food processing: textural comparison of pressure- and heat-induced gels of food proteins. *Agric. Biol. Chem.* 54 (1): 183-189 (1990).

59. K. Gekko, Effects of high pressure on the sol-gel transition of food macromolecules, *High Pressure and Biotechnology* (R. Hayashi, K. Heremans, and P. Masson, eds.), Colloque INSERM John Libbey Eurotext Ltd., 1992, Vol. 224, p. 105.

60. E. A. Elgasim and W. H. Kennick, Effect of pressurization of pre-rigor muscles on protein quality. *J. Food Sci.* 4: 1122-1124 (1980).

61. G. J. Haas, H. E. Prescott, Jr., E. Dudley, R. Dik, C. Hintlian, and L. Keane, Inactivation of microorganisms by carbon dioxide under pressure. *J. Food Safety* 9: 253-265 (1989).

62. A. Carlez, J. P. Rosec, N. Richard, and J. C. Cheftel, Bacterial growth during chilled storage of pressure treated mincemeat. *Lebensm. Wiss. U. Technol.* 27: 48-54 (1994).

63. W. J. Timson and A. J. Short, Resistance of microorganisms to hydrostatic pressure. *Biotechnol. Bioeng.* 7: 139-159 (1965).

64. R. Hayashi and A. Hayashida, Increased amylase digestibility of pressure-treated starch. *Agric. Biol. Chem.* 53: 2543-2544 (1989).

65. J. M. Thevelein, J. A. Van Assche, K. Heremans, and S. Y. Gerlsma, Gelatinization temperature of starch, as influenced by high pressure. *Carbohydrate Res.* 93: 304-307 (1981).

66. S. Ezaki and R. Hayashi, High pressure effects on starch: structural change and retrogradation, *High Pressure and Biotechnology* (R. Hayashi, K. Heremans, and P. Masson, eds.), Colloque INSERM John Libbey Eurotext Ltd., 1992, Vol. 224, p. 163.

67. A. Denda and R. Hayashi, Emulsifying properties of pressure-treated proteins, *High Pressure and Biotechnology* (R. Hayashi, K. Heremans, and P. Masson, eds.), Colloque INSERM John Libbey Eurotext Ltd., 1992, Vol. 224, p. 333.

68. R. Hayashi, Y. Kawamura, and S. Kunugi, Introduction of high pressure to food processing: preferential proteolysis of β-lactoglobulin in milk whey. *J. Food Sci.* 52(4): 1107-1108 (1987).

69. M. Watanbe, T. Makino, K. Kumeno, and S. Arai, High pressure sterilization of ice nucleation-active bacterial cells. *Agric. Biol. Chem.* 55 (1): 291-292 (1991).

3

High Intensity Pulsed Electric Fields: Processing Equipment and Design

I. INTRODUCTION

In this chapter we discuss the application of high intensity pulsed electric fields as one of the nonthermal methods of food preservation. The electric field is applied to the fluid food in the form of short pulses with a pulse duration ranging between a few microseconds and milliseconds. Food may be processed at ambient or refrigerated temperatures. With pulsed electric field treatment, the food is processed in a short period of time and the energy lost due to heating of foods is minimal (1).

Food preservation requires inactivation of the spoilage and pathogenic microorganisms and the enzymes responsible for undesirable reactions in the food. Pulsed electric fields can inactivate the microorganisms and enzymes (2). However, the inactivation occurs when a certain threshold electric field intensity is exceeded. Based on the dielectric rupture theory, the external electric field induces an electric potential difference across the cell membrane known as the *transmembrane potential*. When the transmembrane potential reaches a critical or a threshold value, electroporation or pore formation in the cell membrane occurs. Cell membrane permeability increases as a result of pore formation. The increase in membrane permeability is reversible if the external electric field strength is equal to or slightly exceeds the critical value. The threshold transmembrane potential depends on the specific microorganism or enzyme as well as the medium in which the microorganisms or enzymes are present. For a spherical cell of radius a, the transmembrane potential ΔV is given by (3):

$$\Delta V = 1.5 \, f \, a \, E_o \cos \theta \, \{ 1 - \exp(-t / \tau) \} \tag{1}$$

where f is a constant depending on membrane properties, E_o is the electric field, θ is the angle between the radius vector and the electric field direction, t is the duration of electric field, and τ is the relaxation time given by (3):

$$\tau = f \, a \, C_m \, (r_i + r_e / 2) \tag{2}$$

where C_m is the membrane capacitance per unit area and r_i and r_e are the specific resistances of the internal and external media. The factor f is given by (3):

$$f = 1 / \{ 1 + a \, G_m \, (r_i + r_e) \} \tag{3}$$

where G_m is the membrane conductance per unit area.

For pasteurization purposes, the intensity of the electric field depends on the type of microorganisms or enzymes present in the food. Pulsed electric field inactivation of microorganisms also depends on other factors, including temperature, pH and ionic strength of the food, duration of electric field, and growth stage of the microorganisms (1).

Early studies of electricity for food processing used direct current (DC) or alternating current (AC) electric fields. Later, pulsed electric fields were used in the areas of cell biology and biotechnology for electroporation and electrofusion. Electroporation is the permeabilization of cell membranes under the influence of an electric field for the purpose of gene manipulation or introducing foreign molecules into the cell. Electrofusion is the fusion of cells when subjected to an electric field (1). The application of pulsed electric fields in cell biology is being extended to microbial inactivation for the purpose of food preservation.

II. HIGH INTENSITY PULSED ELECTRIC FIELD PROCESSING SYSTEM

The processing system utilizing high intensity pulsed electric fields consists of a number of components, including a power source, capacitor bank, a switch, treatment chamber, voltage, current and temperature probes, and aseptic packaging equipment (Fig. 1). A power source is used to charge the capacitor bank, and a switch is used to discharge energy from the capacitor bank across the food held in a treatment chamber. A mercury ignitron spark

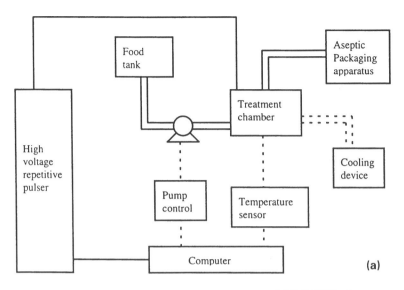

(a)

(b)

FIGURE 1   (a) High intensity pulsed electric field processing system for nonthermal preservation of foods. (From Ref. 1.) (b) Washington State University Pulsed Electric Field Unit. This unit works in continuous mode with a 200 L/h capacity. (c) *Pure Pulse* Pulsed Electric Field concentric cylinder processing chamber [Courtesy of PurePulse Technologies, Inc.]. (d) *Pure Pulse* Pulsed Electric Field processing unit with a coaxial treatment chamber [Courtesy of PurePulse Technologies, Inc.].

(c)

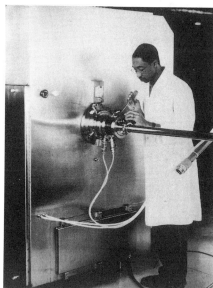

(d)

FIGURE 1   Continued

gap, a gas spark gap, a thyratron, a series of SCRs, a magnet, or a mechanical rotary type switch may be used as the switch (4). Food may be held in a static chamber or may be pumped through a continuous chamber. The static chamber is suitable for preliminary laboratory scale studies. For pilot plant or industrial scale operations, a continuous chamber is preferred. The voltage, current, and electric field strength can be measured using an oscilloscope. The treated food is then filled into individual consumer packages or bulk storage containers using aseptic packaging equipment. It is recommended the processed food be stored at refrigerated temperatures to maintain longer shelf life. The process of subjecting the food to electric fields may be accompanied by the generation of heat. The processing system generally includes a means to provide cooling of the treatment chamber.

One of the important and complicated components of the processing system is the treatment chamber. Several designs of static and continuous chambers are suggested.

III. DESIGN OF STATIC CHAMBERS

A. Sale and Hamilton Chamber

Sale and Hamilton (5) were among the earliest researchers to study the inactivation of microorganisms with high intensity pulsed electric fields. Carbon electrodes supported on brass blocks were selected. A chamber was formed by placing a U-shaped polythene spacer between the electrodes as illustrated in Figure 2. The electrode area and the amount of food that could be treated was varied by using different spacers. The maximum electric field that the chamber could withstand was limited to 30 kV/cm because of the electrical breakdown of air above the food. The temperature of the food was controlled by circulation of water through the brass blocks. Rectangular pulses with a pulse width varying between 2 and 20 µs in steps of 2 µs were selected. The pulse repetition rate was set to 1 pulse per second.

B. Dunn and Pearlman Chamber

The chamber consisted of two stainless steel electrodes and a cylindrical nylon spacer (6). The chamber was 2 cm high with an inner diameter of 10 cm and an electrode area of 78 cm^2 (Fig. 3). The chamber was designed for treating liquid foods. Food was introduced through a small aperture in one of the electrodes. The aperture was also used for temperature measurement during treatment of the food with high intensity electric fields. The electric pulser

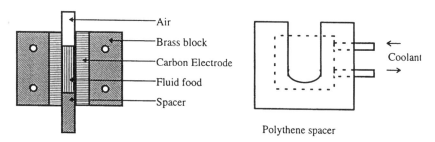

FIGURE 2   Static chamber designed by Sale and Hamilton. (From Ref. 5.)

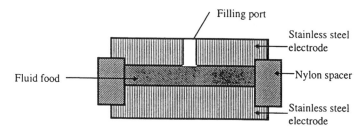

FIGURE 3   Cross section of static chamber designed by Dunn and Pearlman. (From Ref. 6.)

comprised a high voltage power supply, two 400 kΩ resistors, a bank of six
capacitors (each with a capacitance of 0.4 μF), a spark gap switch, a dump re-
lay, a current monitor, and a voltage probe (6).

C.   Grahl et al. Chamber

The treatment chamber designed by Grahl et al. (7) is similar to the Sale and
Hamilton (5) treatment chamber. The electrodes were made of carbon-brass,
and a rectangular Plexiglas™ frame with a thickness of 0.5 or 1.2 cm was used
as the spacer. The effective electrode area was 50 cm^2 and the maximum electric field strength was 30 kV/cm. No cooling was provided for the electrodes
in the chamber.

D.   Washington State University Chamber

The static chamber designed by the Washington State University (WSU)
group consisted of two disk-shaped parallel-plate stainless steel electrodes

separated by a polysulfone spacer (Fig. 4). The effective electrode area was 27 cm^2 and gap between the electrodes was 0.95 or 0.51 cm, using different spacers. The electrodes contained internal jackets to circulate water or a coolant to maintain an acceptable operating temperature (4). The treatment chamber contained two ports for filling and/or withdrawal of the food. The ports also facilitate removal of air in the chamber after the introduction of food into the chamber.

Mean electric field strength, E, is defined as (4):

$$E = V/d \qquad (4)$$

Where V is the electric potential difference between two points in space, and d is the distance between them. The distance, d, for the treatment chamber is equal to the distance between the electrodes. For a uniform electric field, d should be smaller than the diameter of the electrodes.

It is important to take precautions in the design of treatment chambers to avoid dielectric breakdown of foods. Dielectric breakdown occurs when the applied electric field strength exceeds the dielectric strength of the food. The breakdown of food observed as a spark is characterized by (4):

Large electrical current flowing in a narrow channel
Evolution of gas bubbles
Formation of pits on the surface of electrodes
Increase in pressure accompanied by an explosive sound.

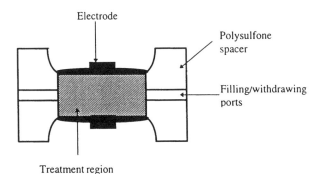

FIGURE 4 Static chamber designed by Washington State University. (From Ref. 4.)

In the WSU chamber design, the surfaces of the electrode are polished to a mirror surface and the edges are rounded to reduce the field enhancement effects and dielectric breakdown of the food.

E. Mizuno and Hori Chamber

Four different types of electrode arrangements were developed to determine if the electrode shape affects microbial inactivation (8).

Plate–Plate

This electrode system consists of two parallel plate electrodes separated by a Plexiglas spacer, which formed the treatment chamber. The chamber length is 10 mm, inner diameter is 8 mm, and volume capacity is 0.5 mL.

Needle–Plate

This electrode system consists of a needle and a plate electrode. The top of the chamber is covered by a Plexiglas plate. A needle electrode is attached to the Plexiglas plate with the tip of the needle protruding 0.5 mm from the surface of the plate. The radius of curvature of the needle tip is about 0.1 mm. A plate electrode is placed at the bottom of the chamber. The needle tip and the plate electrode at the bottom of the chamber are separated by a distance of 9.5 mm. The volume capacity of the chamber is 0.5 mL.

Wire–Cylinder

This electrode system consists of a wire and a cylindrical electrode. The inner diameter of the cylindrical electrode is 19 mm and the wire electrode diameter is 0.5 mm. The wire electrode is held at the center of the cylinder by silicon rubber. The electrode system is 30 mm long with a volume capacity of 8.5 mL.

Rod–Rod

This electrode system consists of rod electrodes in a 6-mm thick polyvinylchloride (PVC) chamber. The rod electrodes are screwed to the chamber at the center, one to the top surface and the other to the bottom surface of the chamber. The tips of the electrodes are separated by a distance of 3 mm. The electrodes are 4 mm in diameter with a 90° cone-shaped tip.

In the described systems, the plate and cylinder electrodes are made of aluminum and the needle, wire, and rod electrodes are made of steel. In the needle–plate, wire–cylinder, and rod–rod electrode systems, the electric field is applied in the form of high voltage arc discharges. Microbial survivability, R, roughly followed the Weibull distribution given by:

$$R = \exp(-\alpha\, N\, W_i) \qquad (5)$$

where α is a constant, N is the number of pulses, and W_i is the input energy. Also the energy input to reduce the survivability by a factor of 10^{-6} differed significantly with the electrode type. The rod–rod, wire–cylinder, needle–plate, and plate–plate electrode arrangements require respectively 5–10, 10–30, 45–90, and 70–370 cal/mL of input energy. The higher efficiency of the needle–plate arrangement compared to the plate–plate arrangement may be attributable to the existence of a region near the needle electrode where the electric field strength is high. Cells distributed in the low electric field region are carried by a convection flow to the high electric field region. However, cells attached to the insulating wall of the chamber, especially to the upper corners of the chamber, may not be carried by the convection flow and thereby may not be inactivated. The wire–cylinder arrangement has still higher efficiency because of the existence of a high electric field region near the wire as well as the absence of insulating wall between the wire and the cylinder electrodes. The rod–rod arrangement has the highest efficiency because electric field is applied in the form of arc discharge.

F. Zheng-Ying and Yan Chamber

Zheng-Ying and Yan (9) developed the rod–rod and three rod electrode systems. The electrodes were placed in a Plexiglas cylindrical chamber with a diameter of 10 mm, length of 50 mm, and volume capacity varying between 1 and 5 mL. Another cylindrical chamber made of PVC pipe with a volume capacity of 1800 mL was similarly designed. Also, a spherical chamber made of copper with paint coating on the inner surface and volume capacity of 1600 mL was designed. The diameter of the electrode was 3 mm with a 60° cone-shaped tip (9). These chambers, however, are not applicable in the food industry because of the electrolysis associated with the arc discharges used for microbial inactivation.

IV. DESIGN OF CONTINUOUS CHAMBERS

A. Dunn and Pearlman Chamber

The processing system designed by Dunn and Pearlman (6) comprises a storage reservoir for the food to be treated, deaeration apparatus, treatment chamber, temperature and voltage monitors, high voltage pulser, and a heat exchanger for preheating the food and cooling the food after the electric field treatment. The preheated food stream is conducted to a deaerator to remove dissolved gases, and subsequently passed through the electric field treatment chamber. The treated food is cooled in the heat exchanger and further cooled in a refrigeration unit to 5° to 10°C. The treated and cooled food is aseptically packed into individual consumer bags or bulk storage containers.

The chamber is made up of two parallel-plate electrodes and a dielectric spacer (Fig. 5). The electrodes are not in direct contact with the food but are separated by ion conductive membranes. Food-grade ion-permeable membranes are made of sulfonated polystyrene, acrylic acid copolymers, or fluorinated hydrocarbon polymers with pendant groups. Electrical conduction between the electrodes and the ion-permeable membranes occurs through an electrolyte. Suitable electrolyte solutions include sodium carbonate, sodium hydroxide, potassium carbonate, and potassium hydroxide. The electrolyte is continuously circulated to remove products of electrolysis. The electrolyte solution is replaced in the event of excess concentration or depletion of ionic components. The residence time of the food in the treatment chamber may be increased by using baffles (6).

Another continuous chamber described by Dunn and Pearlman (6) comprises several electrode reservoir zones separated by dielectric insulating spacers (Fig. 6). The food passes through the openings or orifices between the reservoir zones. The electrodes in each of the reservoir zones may be in direct

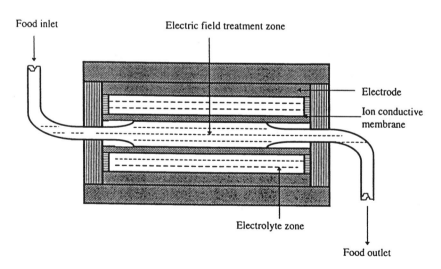

FIGURE 5   Continuous chamber with ion conductive membranes separating the electrodes and the food designed by Dunn and Pearlman. (From Ref. 6.)

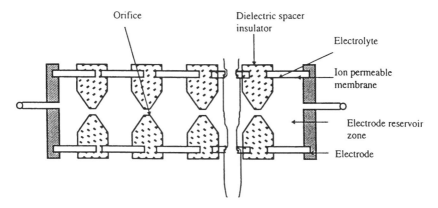

FIGURE 6 Continuous chamber with electrode reservoir zones designed by Dunn and Pearlman. (From Ref. 6.)

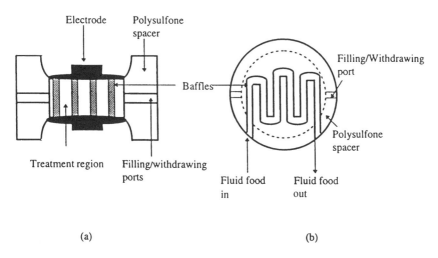

(a)   (b)

FIGURE 7 Continuous chamber with baffles designed by Washington State University. (a) Cross section view, (b) top view. (From Ref. 4.)

contact with the food or may be separated from the food by ion permeable-membranes. The electric field is concentrated in the slot-like openings between the reservoir zones. Liquid food is introduced under high pressure and the average residence time in each of the reservoir zones is less than a minute.

B. Washington State University Continuous Chamber

The WSU static parallel-plate electrode chamber was modified by adding baffled flow channels inside to operate as a continuous chamber (Fig. 7). Two stainless steel disk-shaped electrodes separated by a polysulfone spacer forms the chamber. The designed operating conditions are as follows: chamber volume 20 or 8 mL; electrode gap 0.95 or 0.51 cm; food flow rate 1200 or 6 mL/min (1,4).

V. DESIGN OF A CONTINUOUS COAXIAL CHAMBER

A. Washington State University Chamber

The WSU coaxial chamber was designed based on a modified coaxial cylinder arrangement (Fig. 8). The WSU coaxial chamber uses a protruded outer electrode surface to enhance the electric field within the treatment zone and reduce the field intensity in the remaining portion of the chamber. The electrode configuration was obtained by optimizing the electrode design with a numerical electric field computation. Using the optimized electrode shape a prescribed field distribution along the fluid path without electric field enhancement points was determined. The gap between the electrodes is adjustable in the range of 2 to 6 mm by selecting inner electrodes of different diameters (10). Cooling jackets are built into both electrodes to maintain low temperatures. The design parameters are as follows: outer diameter of the chamber, 12.7 cm; chamber height, 20.3 cm; flow rate, 1 to 2 liters per min (1).

Coaxial configurations can be easily manufactured and give well-defined electric field distributions. The electric field in coaxial chambers is not uniform and depends on the location in the chamber. The field intensity, E, between coaxial electrodes is given by (4):

$$E = V / [r \ln (R_2 / R_1)] \qquad (6)$$

where r is the radius of electric field measurements and R_2 and R_1 are the radii of the outer and inner electrodes, respectively. The uniformity of the electric field in treatment chambers with coaxial electrodes can be improved when

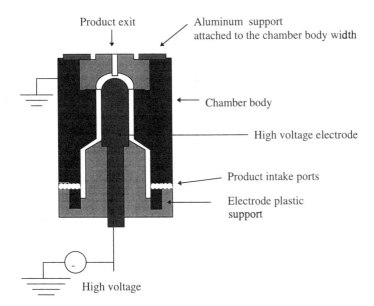

Product exit   Aluminum support
      attached to the chamber body width

Chamber body

High voltage electrode

Product intake ports

Electrode plastic support

High voltage

FIGURE 8   Coaxial cylinder electrode chamber designed by Washington State University. (From Ref. 10.)

$(R_2 - R_1) \ll R_1$. Although the electric field is not uniform, coaxial chambers provide the advantage of a uniform fluid flow and simple chamber structure.

B. Bushnell Continuous Coaxial Chamber

The coaxial chamber comprises an inner cylindrical electrode surrounded by an outer annular cylindrical electrode with the food flowing between the electrodes (11). It is recommended that the length of the treatment chamber not be too small or too long compared to the chamber diameter. If the length of the treatment chamber is too small compared to the diameter, this arrangement results in low efficiency. If, on the other hand, the treatment chamber is very long compared to the diameter, the chamber will have very low electrical resistance for pumpable foods with moderately low resistivity, creating difficulty in the electrical pulser design. The electrical energy, W (joules), consumed in each pulse is given by (11):

$$W = E^2 \, V \, \tau / \, P \qquad (7)$$

where E is electric field (volts/cm),τ is the pulse duration (s), V is the treated volume (cm^3), and P is electrical resistivity of food (ohm-cm).

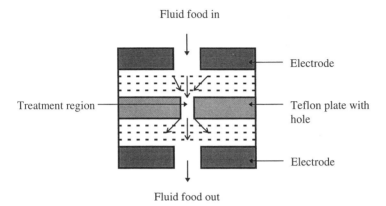

Fluid food in

Treatment region

Electrode

Teflon plate with hole

Electrode

Fluid food out

FIGURE 9 Converged electric field type treatment chamber designed by Matsumoto et al. (From Ref. 12. © 1991 IEEE.)

VI. OTHER CHAMBER DESIGNS

A. Matsumoto et al. Converged Electric Field Type Treatment Chamber

Figure 9 illustrates the converged electric field type treatment chamber designed by Matsumoto et al. (12). Disc-shaped electrodes are separated by 1-cm-thick Teflon™ plates. The electric field is concentrated in a small region. Food is introduced in the concentrated electric field region through a small hole (orifice) in the Teflon plate. Food inside the orifice of the Teflon plate is subjected to the high intensity electric field. The current density at the electrode liquid interface is maintained low to minimize electrolysis and gas bubble formation. This chamber is similar to the continuous chamber with more than one orifice designed by Dunn and Pearlman (6).

VII. GENERATION OF DIFFERENT VOLTAGE WAVEFORMS

Food may be subjected to high intensity electric fields in the form of exponential, square wave, oscillatory, or bipolar pulses. In this section we discuss the generation of these voltage waveforms.

A. Exponential Pulses

In an exponential pulse, the voltage increases to a selected peak value and decreases exponentially. Food subjected to exponential pulses is, therefore, subjected to the peak voltage for a short period of time. Voltage smaller than the peak voltage may or may not have a bactericidal effect in the food. The electrical circuit necessary for generating exponential pulses consists of a DC power supply and a capacitor bank in series with a charging resistor R_c (Fig. 10) (13). The resistance R_1 limits the current in case of an arc discharge in the food, and R_2 controls decay time in the event of food resistivity being greater than expected. The energy density, Q, for exponential decay pulses can be approximated as (4):

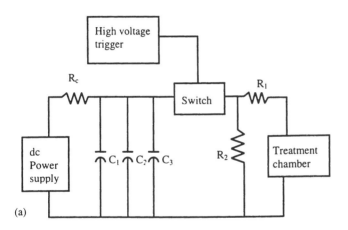

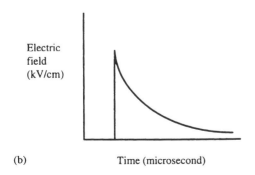

FIGURE 10 (a) Exponential decaying pulse generating circuit. (b) Exponential decaying waveform. (From Ref. 13. © 1994 IEEE.)

$$Q = V_o^2 \, C_o \, n \, / \, 2v = V_o^2 \, t \, / \, 2 \, R \, v \tag{8}$$

where V_o is the initial charging voltage, C_o is the capacitance of the energy storage capacitor, n is the number of pulses, t is the treatment time, R is effective resistance, and v is the volume of treatment chamber.

B.   Square Wave Pulses

A high voltage transmission line connected to a matching load yields a square wave pulse. However, there are two problems in using the high voltage transmission line for square wave pulses. One is the difficulty in matching the resistance of the food with the characteristic impedance of the transmission line (13). The impedance of the transmission line should be matched with the resistance of the food to provide highest energy transfer to the food (11). The other problem is that the high voltage transmission line is not suitable for long pulses because of the high expenses of the long electric cables. These problems can be solved by using a pulse forming network (PFN). The PFN circuit is essentially a distributed transmission line emulated by a number of inductor-capacitor sections (Fig. 11) (13). The energy density, Q, for square wave pulses is given by (4):

$$Q = V \, I \, \tau \, n \, / \, v = V^2 \, \tau \, n \, / \, R \, v = V^2 \, t \, / \, R \, v \tag{9}$$

where V, I, τ are the voltage, current, and pulse width of the square waves, respectively.

C.   Bipolar Pulses

The electrical circuit necessary to generate bipolar pulses is presented in Figure 12 (13). A DC power supply is used to charge the capacitor bank C_1. A signal applied to the series switch SW1 allows a discharge of energy through the capacitor C_2 and the food in the treatment chamber. When the output of the DC power supply is positive with respect to ground, the food is subjected to a positive pulse. When the voltage across the food approaches zero, the shunt switch SW2 is turned on by the controller and energy stored in C_2 is discharged through the food in the form of a negative pulse. A series of bipolar pulses is produced by repeating this sequence. Each time the capacitor C_1 is charged, a bipolar pulse pair is delivered to the food.

D.   Oscillatory Pulses

An oscillatory electric field is generated upon the discharge of a bank of capacitors as presented in Figure 13 (13). An inductor is connected in parallel to

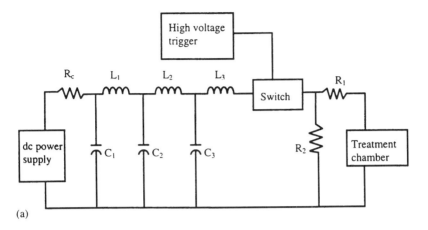

(a)

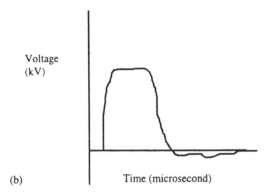

(b)   Time (microsecond)

FIGURE 11   (a) Square wave pulse generating circuit. (b) Square wave-form. (From 13. © 1994 IEEE.)

the treatment chamber. The capacitor bank is charged with a DC power supply. When the switch is closed, an oscillatory voltage is generated across the inductor and the food in the treatment chamber. The oscillatory condition and resonance frequency are determined by the capacitance, inductance, and resistance of the circuit. For the oscillatory voltage to persist for several cycles, the circuit is operated in a lightly damped condition.

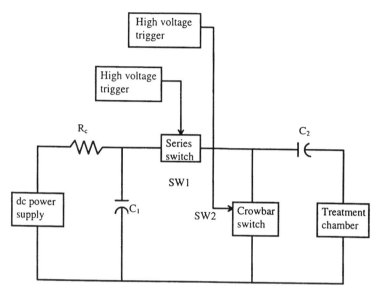

FIGURE 12   Bipolar pulse generating circuit. (From Ref. 13. © 1994 IEEE.)

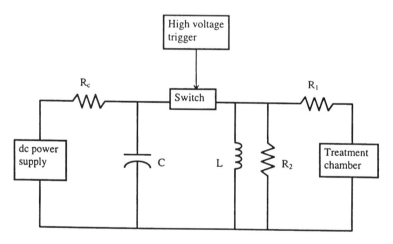

FIGURE 13   Oscillating pulse generating circuit. (From Ref. 13. © 1994 IEEE.)

VIII. SUMMARY

High intensity pulsed electric field processing system is a simple electric system consisting of a high voltage source, a capacitor bank, a switch, and a treatment chamber. Initial studies of high intensity pulsed electric field treatment of foods were conducted using static treatment chambers. Later, continuous and coaxial chambers were designed and developed. The type of electrode arrangement influences microbial inactivation. Other chambers of greater volumes may be designed for use on an industrial scale.

It is important in pulsed electric field processing to avoid the dielectric breakdown of foods. Foods susceptible to dielectric breakdown are not suitable for electric field processing. The risk of dielectric breakdown limits pulsed electric field processing primarily to liquid foods. Although liquid foods with small particulates may potentially be processed, the size of the particulates must be much smaller than the gap of the treatment region. Furthermore, solid foods containing air bubbles are not suitable for electric field processing because air bubbles are potential sites of dielectric breakdown.

REFERENCES

1. B. Qin, U. R. Pothakamury, G. V. Barbosa-Cánovas, and B. G. Swanson, Nonthermal pasteurization of liquid foods using high intensity pulsed electric fields. *Crit. Rev. Food Sci Nutr.* 36: 603–627 (1996).
2. A. J. Castro, G. V. Barbosa-Cánovas, and B. G. Swanson, Microbial inactivation of foods by pulsed electric fields. *J. Food Proc. Pres.* 17: 47-73 (1993).
3. D. Knorr, M. Geulen, T. Grahl, and W. Sitzmann, Food application of high electric field pulses. *Trends Food Sci. Technol.* 5: 71-75 (1994).
4. Q. Zhang, G. V. Barbosa-Cánovas, and B. G. Swanson, Engineering aspects of pulsed electric field pasteurization. *J. Food Eng.* 25: 261-281 (1995).
5. A. J. H. Sale and W. A. Hamilton, Effects of high electric fields on microorganisms I. Killing of bacteria and yeasts. *Biochim. Biophys. Acta* 148: 781-788 (1967).
6. J. E. Dunn and J. S. Pearlman, Methods and apparatus for extending the shelf life of fluid food products. U. S. Patent 4,695,472 (1987).
7. T. Grahl, W. Sitzmann, and H. Märkl, Killing of microorganisms in fluid media by high voltage pulses. *DECHEMA Biotechnol. Conf. Series*, 5B: 675-678 (1992).
8. A. Mizuno and Y. Hori, Destruction of living cells by pulsed high voltage application. *IEEE Trans. Ind. Appl.* 24(3): 387-394 (1988).
9. L. Zheng-Ying and W. Yan, Effects of high voltage pulse discharges on microorganisms dispersed in liquid. *Eighth International Symp. High Voltage Eng.*, Yokohoma, Japan, Aug 23-27 (1993).

10. B. Qin, Q. Zhang, G. V. Barbosa-Cánovas, B. G. Swanson, and P. D. Pedrow, Pulsed electric field treatment chamber design for liquid food pasteurization using finite element method. *Trans. ASAE* 38 (2): 557-565 (1995).

11. A. H. Bushnell, J. E. Dunn, and R. W. Clark, High pulsed voltage systems for extending the shelf life of pumpable food products. U. S. Patent 5, 048, 404 (1991).

12. Y. Matsumoto, T. Satake, N. Shioji, and A. Sakuma, Inactivation of microorganisms by pulsed high voltage applications. *Conf. Rec. IEEE Industrial Appl. Soc. Ann. Meeting*, pp. 652-659 (1991).

13. B. Qin, Q. Zhang, G. V. Barbosa-Cánovas, B. G. Swanson, and P. D. Pedrow, Inactivation of microorganisms by pulsed electric field of different voltage waveforms. *IEEE Trans. Dielectrics Electrical Insulation* 1 (6): 1047-1057 (1994).

4

Biological Effects and Applications of Pulsed Electric Fields for the Preservation of Foods

I. INTRODUCTION

In the preceeding chapter, we discussed the equipment used in the application of electric fields to foods. We now turn to the effects of electric fields on microorganisms, enzymes, chemical reactions, and the sensory quality of food as well as the potential application of pulsed electric fields (PEF) as a nonthermal method of food preservation.

Milk pasteurized by using electric fields does not possess the "objectionable features" of thermally pasteurized milk (1). In 1935, Getchell described an electric pasteurization facility for milk, with a capacity of 100 gallons per hour, as a simple, flexible, economical, fast, and dependable operation for preserving milk. The passage of electric current through milk generated heat responsible for the inactivation of microorganisms. However, electric field pasteurization did not become popular for two reasons: (a) the inability to develop equipment with the provision for controlling the temperature of the milk, and (b) the lack of capacity to handle large volumes of milk. Although both electric and thermal pasteurization methods can inactivate microorganisms, the quality of thermally pasteurized milk, in terms of natural "fresh-like" characteristics, is poor compared to electrically pasteurized milk.

If electric fields are applied to the food product in the form of short-duration pulses rather than passing electricity directly through the food product, minimal heat is generated and the process remains nonthermal. The pulse repetition rate also contributes to heat generation.

II. PULSED ELECTRIC FIELDS AND MICROORGANISMS

A. Pulsed Electric Fields and Microorganisms in Model Foods

Sale and Hamilton (2) were pioneers conducting a systematic study to assess the nonthermal bactericidal effects induced by electric fields. Prior to the studies conducted by Sale and Hamilton (2), some reports indicated bactericidal effects; others indicated mutations in microorganisms due to pulsed electric fields. Thus, the study of the effects of electric fields on microorganisms was inconclusive. Castro et al. (3) reviewed the studies on electric field pulses for inactivation of microorganisms and also discussed the mechanisms of microbial inactivation induced by electric fields and processes such as Electro-Pure, Elsteril, Elcrack, and Bach.

Most of the early research on the bactericidal effects of pulsed electric fields was conducted using model food systems such as distilled water, deionized water, phosphate buffer, simulated milk ultrafiltrate (SMUF), and other media (2,4–10). The model systems minimize the interfering factors in terms of the constituents of the medium. The trends of electric fields in model systems can be extended to "real" foods containing fats, proteins, carbohydrates, and other constituents.

Sale and Hamilton (2) exposed microorganisms including *Escherichia coli, Micrococcus lysodiekticus, Sarcina lutea, Bacillus subtilis, Bacillus cereus, Bacillus megaterium, Clostridium welchii, Saccharomyces cerevisiae,* and *Candida utilis* suspended in neutral sodium chloride solution to an electric field intensities between 5 and 25 kV/cm. The electric field was applied in the form of short rectangular pulses with a pulse duration between 2 and 20 μs. The pulse repetition rate was 1 pulse per second. The long interval between pulses helps to minimize an increase in temperature. *Saccharomyces cerevisiae* was most sensitive and *Micrococcus lysodiekticus* was most resistant to the electric field treatment (Fig. 1). Although electrolysis of the medium occurred during the treatment, inactivation of microorganisms is not caused by the electrolytic products. Electrolysis was observed as a burst of nascent hydrogen gas generated at the electrode surface by each pulse. The gas, however, did not penetrate a gel inoculated with *E. coli* and placed in the treatment chamber. The gel was composed of nutrient agar and 10% rezazurin. Inactivation of *E. coli* occurred on the surface of the gel in contact with the cathode as well as in the interior parts of the gel. Therefore, inactivation of *E. coli* was not caused by electrolysis. Hamilton and Sale (4) demonstrated a two-log reduction of *E. coli* in 0.1% saline and *S. aureus* in phosphate buffer (20 mM, pH 7.2).

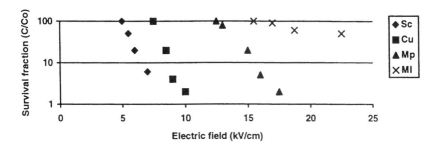

FIGURE 1 Relationship between inactivation and electric field treatment of various microorganisms subjected to 10 pulses of 20 μs pulse duration. Sc, *Saccharomyces cerevisae*; Cu, *Candida utilis*; Mp, motile *Pseudomonad*; Ml, *Micrococcus lysodiekticus*. (From Ref. 2.)

$E. coli$ cells suspended in NaCl (17.4 mM), $Na_2S_2O_3$ (8.83 mM), or NaH_2PO_4/Na_2HPO_4 (7.44 mM) were inactivated 99.9% after 10 pulses at 20 kV/cm. Inactivation of $E. coli$ in sulfate and phosphate solutions is lower than in the chloride solution when the initial cell concentration is low (10^5 cells/mL). With a high initial concentration (10^8 cells/mL) of $E. coli$, inactivation is similar in chloride, sulfate, and phosphate solutions (5). Inactivation in pulse-treated chloride solutions may be the result of the electrolytic production of free active chlorine created by the anodal oxidation of chloride ions. Hypochloric acid, the bactericidal agent, is produced in a secondary step from a reaction of chlorine with water (5).

Pothakamury et al. (6,7) reported a four to five log cycle reduction of *Lactobacillus delbrueckii* (ATCC 11842), *B. subtilis* (ATCC 9372), *S. aureus* (ATCC 6538), and *E. coli* (ATCC 11229) in SMUF after 40, 50, and 60 pulses, respectively at 16 kV/cm (Figs. 2 through 5). The treatment volume used in these studies was 0.1 mL.

Lactobacillus brevis cells suspended in a phosphate buffer solution, $Na_2HPO_4/NaH_2PO_4 \cdot H_2O$ (0.845/0.186 mM), and exposed to an electric field of 30 kV/cm at 24°C were reduced by six log cycles after 3.6 ms. Increasing the temperature to 60°C resulted in a nine log cycle reduction after 10 ms with an electric field of 25 kV/cm (8). The inactivation at 60°C may be partially caused by thermal effects.

Application of 20 pulses at 25 kV/cm and 25°C to *E. coli* (ATCC 11229) suspended in SMUF resulted in a 2.8 log cycle reduction (Fig. 6). On the other hand, *S. cerevisiae* (ATCC 16664) was reduced by 3.9 log cycles after 18 pulses at 25 kV/cm and 25°C (Fig. 7). A greater degree of inactivation

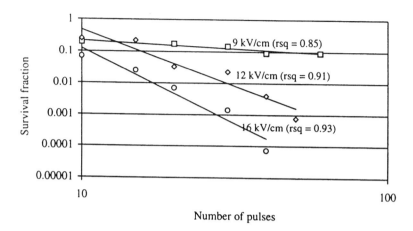

FIGURE 2 Inactivation of *Lactobacillus delbrueckii* subjected to 9 (□), 12 (◇) and 16 (○) kV/cm. (From Ref. 7.)

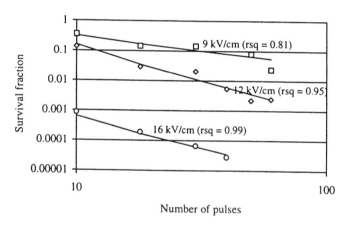

FIGURE 3 Inactivation of vegetative cells of *Bacillus subtilis* subjected to 9 (□), 12 (◇) and 16 (○) kV/cm. Each experimental condition was tested in duplicate and each point in the graph is an average of three points. (From Ref. 7.)

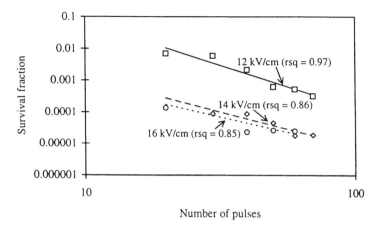

FIGURE 4 Inactivation of *E. coli* in SMUF after 20, 30, 40, 50, and 60 pulses using 12 (□), 14 (◇), and 16 (○) kV/cm. r^2 was calculated and found to be as follows: 12 kV/cm, 0.97; 14 kV/cm, 0.86; 16 kV/cm 0.85. (From Ref. 6.)

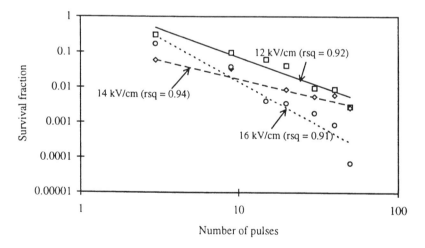

FIGURE 5 Inactivation of S. aureus in SMUF after 20, 30, 40, 50, and 60 pulses using 12 (□), 14 (◇), and 16 (○) kV/cm. r^2 was calculated and found to be as follows: 12 kV/cm, 0.92; 14 kV/cm, 0.94; 16 kV/cm 0.91. (From Ref. 6.)

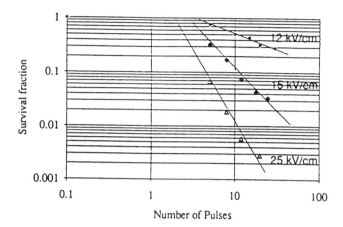

FIGURE 6   Survival rate of *E. coli* vs. number of pulses at electric fields of 12, 15, and 25 kV/cm. The capacitance used was 1.8 μF. The energy input corresponding to those electric fields were 130, 212, 604 J respectively. Each data point corresponds to one treated sample. (From Ref. 9.)

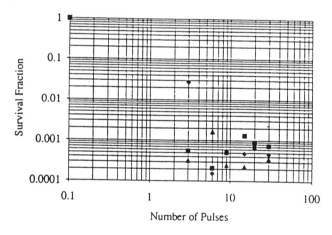

FIGURE 7   Survival rate of *S. cerevisiae* vs. number of pulses at electric fields of 12 (♦), 15 (■), and 25 (▲) kV/cm. The capacitance used was 1.8 μF. The energy input corresponding to those electric fields were 120, 196, 558 J respectively. Each data point corresponds to one treated sample. (From Ref. 9.)

is obtained with a reduction in the initial concentration of *S. cerevisiae*. However inactivation is not affected by the initial concentration of *E. coli* in the treatment sample (9).

A stepwise treatment approach was adopted by Zhang et al. (10) to increase the inactivation of *E. coli* (ATCC 11229) in SMUF. Using a static chamber, five steps of 80 pulses (16 pulses/step) at 40 kV/cm were used. In the first step, eight 14-mL batches were treated with 16 pulses. In the second step, the treated batches from the first step were mixed and again divided into 14-mL batches. Each 14-mL batch was subjected to 16 pulses at 40 kV/cm. The third, fourth, and fifth steps were similar to the second step. After five steps, the *E. coli* population was reduced by nearly nine log cycles (Fig. 8). The energy required to achieve the nine-log reductions was 97 kJ/L.

Most researchers use a pumpable fluid for studying the effects of pulsed electric fields. Because not all foods are pumpable, it is also necessary to know the potential of PEF for preserving semisolid and solid food products. The inactivation of *E. coli* (ATCC 11229), *S. aureus* (ATCC 6538), and *S. cerevisiae* (ATCC 16664) in potato dextrose agar, a model semisolid system, was studied by Zhang et al. (11). *E. coli* and *S. aureus* were suspended in SMUF, and *S. cerevisiae* was suspended in apple juice. The cell suspensions were serially diluted in 0.1% peptone. Sterilized potato dextrose agar, while still liquid, was inoculated with one milliliter of the peptone suspension. The inoculated PDA was filled in a 14 mL static chamber and allowed to solidify. The solidified PDA containing the microorganisms was treated with electric pulses. *E. coli* and *S. aureus* were reduced by six log cycles after 64 pulses at 40 kV/cm. Only 16 pulses were required to reduce the *S. cerevisiae* population by six log cycles (Fig. 9).

B. Pulsed Electric Fields and Microorganisms in "Real" Foods

Several reports present promising PEF-induced inactivation of microorganisms inoculated into buffered or model food systems. The challenge is to apply pulsed electric field technology to process foods in the real world. Furthermore, the potential of PEF technology can progress only if PEF offers food products of better quality than thermally processed products. The challenge of preserving "real" foods with PEF was undertaken at Washington State University through the processing of food products such as apple juice, eggs, milk, and pea soup with pulsed electric fields.

Fresh apple juice (Tree Top Inc., Selah, Wash.) and juice reconstituted from concentrate were processed. The shelf life of the PEF-treated juice

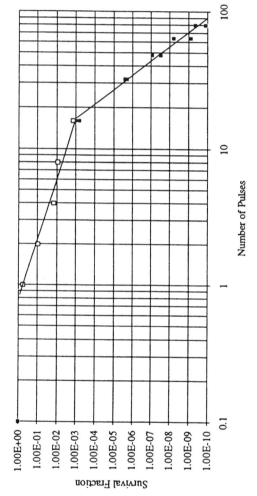

FIGURE 8   Inactivation of *E. coli* with high strength short duration pulsed electric fields. Electric field intensity was 70 kV/cm; pulse width 2 μs; temperature 20°C. (From Ref. 10.)

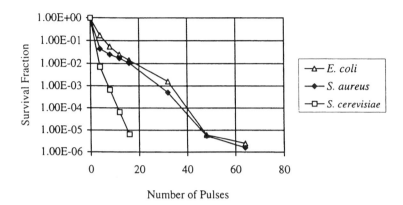

FIGURE 9 Inactivation of *Escherichia coli, Staphylococcus aureus*, and *Saccharomyces cerevisae* in potato dextrose agar in relation to the number of pulses of 3 μs duration. Electric field intensity was 40 kV/cm and temperature 15±1°C. (From Ref. 11.)

stored at 4°C was 3 to 4 weeks. The stored product was analyzed to determine if pulsed electric fields modified the chemical composition of the juice. Under the experimental conditions (maximum of 40 kV/cm) used in the study, the total solids concentration of 11% consisting of 10% carbohydrates, 0.2% ash, and traces of fat and protein was equivalent before and after the electric field treatment. The pH varied between 4.1 and 4.4 and was not significantly affected by the electric field treatment. Vitamin C (ascorbic acid) concentration remained unaffected by the electric field treatment. The concentrations of Ca, Mg, Na, and K were significantly reduced in the PEF-treated apple juice (12). A sensory acceptance panel observed no significant differences between the treated and untreated juices.

Thirty-five pulses of an electric field of ~35 kV/cm applied to freshly produced high pulp orange juice resulted in over a five-log reduction of the naturally occurring microbiological contaminants. The color and taste of the juice was acceptable for at least 10 days. Untreated orange juice was unacceptable after 4 days (13).

Milk with 2% fat was treated with PEF, aseptically filled into sterile pouches, and stored at 4°C for shelf-life studies. Unlike apple juice, electric field–treated milk exhibited a shelf life of 2 weeks. Electric field treatment did not affect the physical or chemical properties of milk. The sensory evaluations indicated no differences between the heat-pasteurized and electric field–treated milk (14).

E. coli (ATCC 10536) inoculated in homogenized and pasteurized milk was 99.9% inactivated when subjected to 23 pulses at ~ 43 kV/cm. *Salmonella dublin* (California State Health Laboratory) inoculated in homogenized and pasteurized milk (3,800 CFU/mL), was inactivated completely after 40 pulses at 36.7 kV/cm. The PEF treatment reduced the population of other bacteria in the milk to less than 20 CFU/mL. No growth of *S. dublin* occurred up to 8 days in the PEF-treated milk stored at 7–9°C. However the population of the other bacteria increased to 400 CFU/mL (13).

The shelf life of fluid eggs processed with pulsed electric fields and aseptically filled in sterile pouches is 4 weeks at 4°C (14). Dunn and Pearlman (13) reported a 4-week shelf life at 4°C for fluid eggs treated with pulsed electric fields in combination with additives such as potassium sorbate and citric acid.

Although the chemical properties of the eggs are not affected by the electric fields, the viscosity of the treated eggs decreases and the color darkens. No significant differences in the sensory evaluation were observed between a scrambled preparation of fresh and electric field–treated eggs. Furthermore, the electric field–treated scrambled eggs were preferred to the scrambled eggs prepared from a commercial preserved liquid egg (14).

Green pea soup consisting of split pea powder, starch, hickory smoke flavor, granulated sugar, monosodium glutamate, and distilled water was preprocessed by mixing and heating. The soup was subjected to pulsed electric field treatment. The shelf life of the pea soup is about 4 weeks at 4°C. The chemical and physical properties of the pea soup did not change after the PEF treatment and during storage. Pulsed electric field treatment did not alter the sensory properties of the pea soup (14).

Electric fields inactivate *S. cerevisiae* more readily than Lactobacilli inoculated into yogurt. Based on the different inactivations, the shelf life of yogurt inoculated with the yeast and treated with electric fields at 45°C can be increased by 10 days when stored at 4°C. Increasing the PEF treatment temperature to 55°C increases the shelf life at 4°C by one month (13).

C. Factors Affecting Microbial Inactivation by PEF

Microbial inactivation increases with an increase in the electric field strength, number of pulses, pulse duration, pulse shape, temperature of the medium, maturity of the bacteria, and ionic strength of the medium (18).

Inactivation increases greatly when the field strength E exceeds a critical value E_c (Fig. 10). The survival rate, s, is related to the electric field strength, E, according to the equation given by (19):

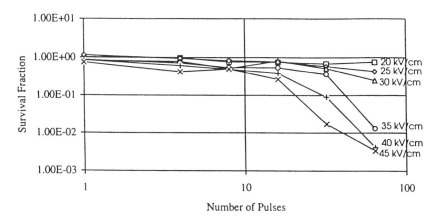

FIGURE 10   Inactivation of *E. coli* in skim milk as a function of electric field intensity using WSU static chamber. (From Ref. 18.)

$$\ln s = - b_E (E - E_c) \tag{1}$$

where b_E is the regression coefficient, and E_c is the extrapolated critical value of E for 100% survival. Increasing the electrical resistivity of the suspension changes the threshold value E_c and increases inactivation. The survival rate may also be related to the time of treatment, t, as given by (19):

$$\ln s = -b_t \ln\left(\frac{t}{t_c}\right) \tag{2}$$

where b_t is the regression coefficient, and t_c is the extrapolated value of t for 100% survival. The survival rate as a function of electric field strength and treatment time is given by:

$$s = \left(\frac{t}{t_c}\right)^{-(E-E_c)/k} \tag{3}$$

The parameters E_c, t_c, and k are specific to the microorganism. The model given by equation 3 is influenced by suspension temperature and bacterial cell concentration (19). The survival rate decreases or, in other words, the inactivation increases with an increase in the electric field strength and treatment time.

The time of treatment is the product of number of pulses and pulse duration. Increasing the treatment time means an increase in either the pulse duration or the number of pulses. Increasing the pulse duration may be accompanied by a large increase in temperature of the treatment system. Therefore, the pulse duration can be increased to a value so the increase in temperature is acceptable.

Pulsed electric fields may be applied as exponential decay, square, wave, or oscillatory pulses. The pulses may be monopolar or bipolar. The effectiveness of the pulses to inactivate microorganisms decreases in the order of square wave, exponential decay, and oscillatory. At a pulse number less than 20, square wave pulses result in 60% more inactivation than exponential decay pulses applied to *S. cerevisiae* in an electric field of 12 kV/cm. However, beyond 20 pulses, microbial inactivation is similar with square wave and exponential decay pulses (Fig. 11). The energy efficiency of square wave pulses is calculated to be 91%; exponential pulses are 64% energy efficient (20).

Bipolar pulses are more lethal than monopolar pulses (Fig. 12). A monopolar pulse has a exponential decay pulse waveform; a bipolar pulse consists of one positive and one negative exponential decay waveform. Each application of a bipolar pulse is equivalent to applying two monopolar pulses.

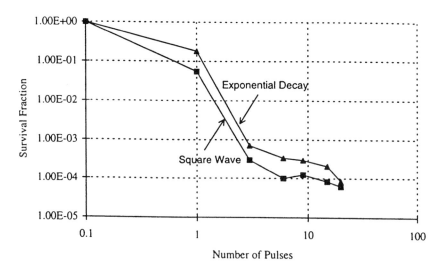

FIGURE 11 Survival fractions of *S. cerevisiae* after pulsed electric field treatment using exponential and square wave pulses. Electric field intensity was 12 kV/cm, pulse width 90 μs for exponential and 60 μs for square wave pulses. (From Ref. 20.)

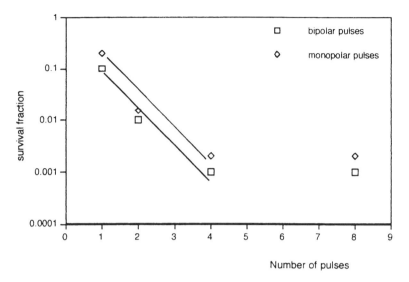

FIGURE 12   Survival fraction of *E. coli* after treatment with monopolar and bipolar exponential decay waveforms. Peak electric field was 40 kV/cm and applied energy 60 J/pulse for both the waveforms. (From Ref. 21. © 1994 IEEE.)

Application of bipolar pulses causes a reversal of the electric charge after every pulse. Charge reversal changes the direction of movement of charged ions in the cell membrane. The amplitude of movement of the ions may not be large enough to result in a mechanical breakdown of the cell membrane but certainly causes a structural fatigue in the membrane and increased susceptibility to electrical breakdown. Due to the reversible reactions at the electrodes, bipolar pulses reduce the undesirable electrolysis of the liquid food being treated. Monopolar pulses separate charged particles in liquid foods. The separated charged particles may form a deposit on the electrode and distort the electric field. Bipolar pulses do not separate charged particles and therefore no deposition of the particles occurs (21).

According to Sale and Hamilton (2), the growth phase of microorganisms and temperature of the medium do not affect the inactivation. However, other researchers observed that inactivation is a function of microorganism growth phase and medium temperature (18,22,23). Logarithmic phase cells are more sensitive to electric fields than are stationary phase cells. Hülsheger et al. (22) demonstrated greater sensitivity of a 4-h culture of *E. coli* to pulsed electric fields compared with a 30-h culture (Fig. 13). Similar results were ob-

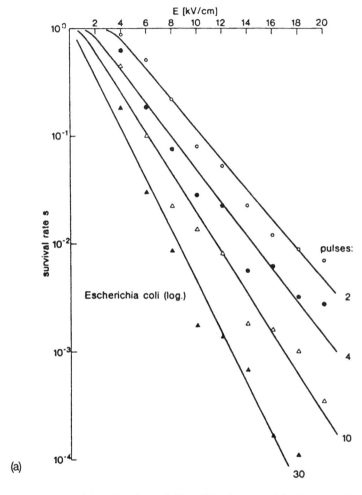

(a)

FIGURE 13   (a) Inactivation of *E. coli* in the logarithmic growth phase (4 h culture). (b) Inactivation of *E. coli* in the stationary growth phase (30 h culture). (From Ref. 22.)

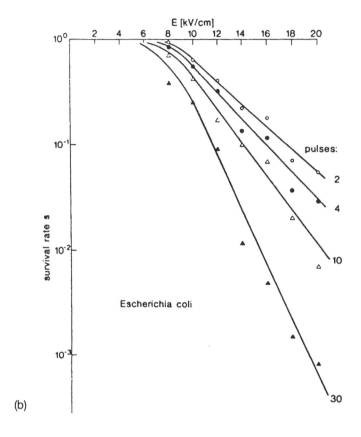

(b)

tained by applying two or four pulses at 36 kV/cm to *E. coli* cells suspended in SMUF (Fig. 14). The logarithmic cells of *S. cerevisiae* are more sensitive to pulsed electric fields than are cells in the stationary phase. The growth area, incorporating separation of mother and daughter yeast cells during budding, is especially sensitive to the electric field treatment (3).

Inactivation increases with an increase in temperature of the medium. With exponential decay pulses, population of *E. coli* in SMUF reduces by two log cycles after 20 pulses at 40°C and 50 pulses at 30°C. On the other hand, when square wave pulses are used, population of *E. coli* in SMUF reduces by two log cycles after 10 pulses at 33°C and 60 pulses at 7°C (23).

Inactivation increases with a decrease in the ionic strength of the medium. After 30 pulses at 40 kV/cm, the population of *E. coli* in SMUF decreases by two log cycles with a decrease in the ionic strength from 0.0168 M

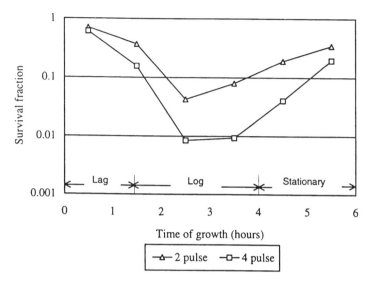

FIGURE 14 Inactivation of *E. coli* harvested in the lag, logarithmic, and stationary growth phases. Harvested cells were suspended in simulated milk ultrafiltrate and subjected to 2 and 4 pulses at 36 kV/cm. (From Ref. 23.)

to 0.028 M. There is a slight increase in the inactivation of *E. coli* in SMUF with a decrease in pH (24). Microbial inactivation is not influenced by the presence or absence of oxygen during the growth of *E. coli* (2).

The presence of Na or K in the treatment medium does not affect microbial inactivation, whereas bivalent cations Mg and Ca induce a protective mechanism against electric field treatment (19).

III. PULSED ELECTRIC FIELDS AND ERYTHROCYTES

Exposure of erythrocytes to electric fields of 22 kV/cm resulted in a 99% inactivation of cells because of cell lysis. The electric field–induced cell lysis is similar to lysis resulting from suspension of cells into a hypertonic medium. The cell membrane is intact, but loses the property of semipermeability. However, the bimolecular structure of the membrane is not affected by the application of electric fields (4).

The application of pulsed electric fields can promote many different chemical reactions and physical effects in erythrocytes, e.g., lipid thermal phase transition, electrophoretic movement of ions or charged molecules, and electroconformation changes in the bulk solution and in the cell membrane. The primary effect of electric fields is to create aqueous pores of limited size into cell membranes. With a single square wave DC pulse of approximately 4 kV/cm, 5–120 μs, the electropores are large enough to allow permeation of ions and small molecules (molar mass < 2000 g), but not the passage of hemoglobin or enzymes. Approximately 200 pores with a mean size of 1–3 nm are created per cell. The pores have a lifetime of one second to a few minutes after the electric pulse is turned off. The size of the pores depends on the electric field intensity, pulse duration, and ionic strength of the medium. A higher intensity, longer pulse, or lower ionic strength of the suspending medium leads to the formation of larger pores. If the cells are subjected to electric pulses with an intensity much stronger than the critical strength, and longer than 10 ms, mechanical breakdown of the membrane may occur. A large piece of membrane may also be damaged and broken apart from the cell (15).

IV. PULSED ELECTRIC FIELDS AND BACTERIAL SPORES

Although the vegetative bacteria were inactivated by electric fields, spores of *B. cereus* and *B. polymyxa* were resistant even in an electric field of 30 kV/cm. However, the spores are sensitive to electric fields after germination (4). Electric field pulses do not induce germination of spores and therefore do not inactivate spores. However, if germination of spores can be induced by other methods, electric field pulses can be used to inactivate the resulting vegetative cells (16). Simpson et al. (17) report nearly a five-log reduction of *B. subtilis* spores using a combination of lysozyme and electric field pulses. The lysozyme probably dissolves the spore coat and renders the cell susceptible to the electric field. Thus, inactivation of spores can be achieved by combining electric field technology with other methods.

V. MECHANISMS OF PEF MICROBIAL INACTIVATION

Electrical breakdown of biological membranes was extensively explored based on model systems such as liposomes, planar bilayers, and phospholipid vesicles because lipid bilayer membranes exhibit properties similar to those of cell membranes (25). The breakdown of lipid bilayers and biological cell membranes can be explained based on (a) dielectric breakdown, (b) threshold transmembrane potential and compression of the cell membrane, (c) viscoelastic properties of cell membranes (26), (d) fluid mosaic arrangement of lipids and proteins in the cell membrane (27), (e) structural defects in the membranes (28), and (f) colloidal osmotic swelling (15).

Zimmermann et al. (29) suggested that exposure of biological cells to electric fields causes an ionic punch-through effect and dielectric breakdown of the cell membrane. The ionic punch-through effect is the rapid increase in the membrane conductance coupled with ionic movement. The change in membrane structure and permeability observed at the critical breakdown voltage is called dielectric breakdown of the cell. In red blood cells and bacteria, dielectric breakdown of the cell membranes can be induced by the application of electric field strengths on the order of 1000 to 10,000 V/cm.

When exposed to an intense electric field, the bilayers of phospholipid vesicles are polarized because of the movement of ions along the electric field lines. Because the lipid bilayer is a poor conducting medium, the ions accumulate at the surface of the bilayer and generate a transmembrane potential (30,31). When the induced transmembrane potential is greater than the natural potential of (~1 V) the cell, membrane rupture takes place. The induced transmembrane potential is a function of the electric field intensity and the cell size. For the assumption of spherical cells surrounded by non-conducting membranes, the induced potential, V_m, is given by (22):

$$V_m = f \, a \, E_c \tag{4}$$

where f is the form factor equal to 1.5 for spherical cells, a is the cell radius, and E_c is the applied electric field intensity. The induced membrane potential for nonspherical cells is determined based on the assumption that the cell shape consists of a cylinder with two hemispheres on each end (22).

The electric potential, V_m, across the cell membrane based on the viscoelastic properties of the membrane is given by:

$$V_m = \left(\frac{24 \, \sigma \, G \, h^3}{\varepsilon_m^{\,2} \varepsilon_o^{\,2}} \right)^{0.25} \tag{5}$$

where σ is surface tension, G is membrane shear elasticity modulus, h is membrane thickness, ε_m is membrane relative permittivity, and ε_o is equal to 8.85×10^{-12} F/m (26).

Cell breakdown may also be explained based on the compression of membranes under intense electric fields. In a simple model of the cell, the cell membrane can be regarded as a capacitor filled with a dielectric material of a low dielectric constant (Fig. 15a). The typical dielectric constant of the bi-

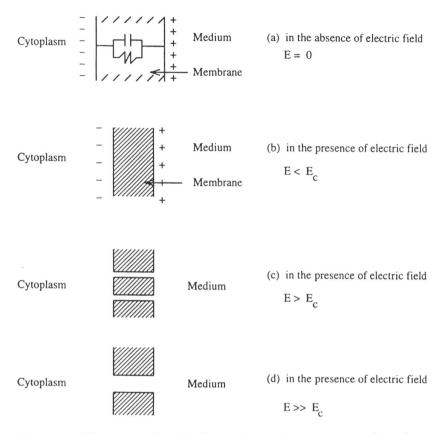

FIGURE 15 Electroporation of cell membrane due to compression when exposed to high intensity electric fields. E_c represents the critical electric field. (From Ref. 18.)

layer structure of a cell membrane is 2. The dielectric constant inside the cell is much greater than the dielectric constant of the cell membrane. If the dielectric constant of the medium containing the cells is high, free charges can accumulate on both sides of the membrane surfaces. Application of an external electric field causes accumulation of additional surface charges and an increase in the potential difference across the membrane. The accumulated charges at the two membrane surfaces are opposite and attract each other, resulting in membrane compression (Fig. 15b). The electro-compressive force, P_e, per unit area of membrane is given by (27):

$$P_e = \frac{-d}{d\delta} \int_o^\delta \frac{1}{2} \ \varepsilon \ \varepsilon_o E^2 dx \qquad (6)$$

where δ is the thickness of the membrane, E is the electric field, ε is the dielectric constant or relative electric permittivity, and ε_o is the electric permittivity of free space.

If the compressive force is independent of the position, x, then

$$P_e = \frac{\varepsilon \ \varepsilon_o V^2}{2\delta^2} \qquad (7)$$

where V is the electric potential difference across the membrane. The compression of the membrane creates elastic strain forces. Assuming that the membrane is an ideal (linear) elastic material, the mechanical restoring force P_m per unit area is given by (27):

$$P_m = Y \ln \frac{\delta}{\delta_o} \qquad (8)$$

where Y is the elastic, compressive modulus of the membrane, defined as the rate at which the thickness decreases with pressure, and δ_o is the unstrained membrane thickness.

For equilibrium: $P_e + P_m = 0$ (9)

For small values of δ, P_e will increase more rapidly than P_m with decreasing δ, and the membrane is not likely to have linear elastic properties. When P_e exceeds P_m, membrane breakdown or pore formation occurs (Fig. 15c). The threshold or critical electric field is the electric field intensity that results in membrane breakdown. The critical potential difference for electromechanical breakdown of the membrane, V_c, is given by (27):

$$V_c^2 = \frac{0.3679 \ Y \ S^2}{\varepsilon \ \varepsilon_o} \qquad (10)$$

With an increase in exposure times and electric field intensities beyond the threshold, larger and larger areas of the membrane are subjected to breakdown (Fig. 15d). If the size and number of pores become large in relation to the total membrane surface, membrane breakdown results in physical destruction of the cell (3,32).

Inactivation of *Valonia utricularis* by PEF increases with an increase in temperature. Inactivation increases because the critical potential difference decreases with increasing temperature, membrane resistance decreases with increasing temperature, and the membrane elastic modulus for deformation normal to the plane of the membrane decreases with an increase in temperature (27).

Dimitrov (26) explained the electric field–induced membrane breakdown based on viscoelastic membrane properties. Membrane breakdown is assumed to occur in three stages: (a) growth of membrane surface "fluctuations," (b) molecular rearrangements leading to membrane discontinuities, and (c) expansion of pores resulting in a mechanical breakdown of the cell (Fig. 16).

An alternative mechanism for the electrical breakdown of membranes is based on the fluid mosaic model of lipids and proteins in the cell membranes (27). In the fluid mosaic model, the membrane is assumed to contain protein units embedded in lipid bilayers (Fig. 17). In the absence of an electric field, the protein units do not form channels through the membrane. Exposure to an electric field causes compression of the membrane, and the protein units form a conducting channel through the membrane, thereby increasing membrane permeability.

Irreversible damage of the lipid bilayer is also explained on the basis of structural defects in the membrane. When the applied high intensity electric field exceeds the threshold transmembrane potential, the spontaneous structural irregularities existing in the membrane develop into pores. The breakdown of cell membranes can also be explained on the basis of structural

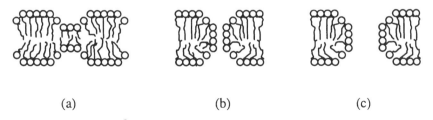

(a)   (b)   (c)

FIGURE 16   Electroporation of cell membranes based on their viscoelastic properties. (From Ref. 26.)

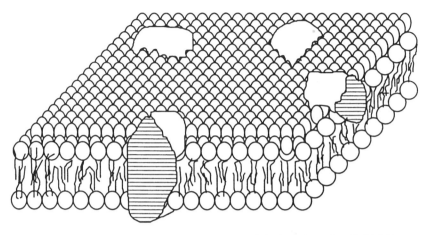

FIGURE 17 A fluid mosaic membrane model showing the lipid bilayer and protein units embedded in the bilayer. (From Ref. 37.)

defects, but the threshold potential of lipid domains in the cell membranes are probably much smaller than artificially prepared lipid bilayers (31).

Serpersu et al. (33) described the damage of human erythrocyte membranes exposed to high intensity electric fields based on colloidal osmotic swelling of the cell. An electrically perforated cell membrane becomes permeable to ions and molecules of a chemical substance used as a probe. The pores are small enough to block the passage of hemoglobin. As a result, the colloidal osmotic pressure of hemoglobin will lead to an influx of water. The cell swells due to water influx, and when the cell volume approaches about 155% of the original volume in the case of human erythrocytes, the swelling leads to membrane rupture and release of the cytoplasmic contents (15,34,35). The process of a colloidal osmotic swelling of a cell is illustrated in Figure 18.

PEF induces pore formation in bilayer lipid membranes in a two-stage process (28): (a) formation of hydrophobic pores and (b) formation of hydrophilic pores (Fig. 19). Hydrophobic pores are formed by the thermal motion of lipid molecules in the lipid bilayer. The lifetime of the hydrophobic pores is on the order of lipid molecule fluctuations. A pore formed may be hydrophobic or hydrophilic, depending on pore energies at different pore radii. Pore energy is the change of free energy resulting from the formation of a cylindrical pore of radius r in the lipid bilayer. A hydrophobic pore of zero radius with zero energy represents the membrane in an undisturbed state. When

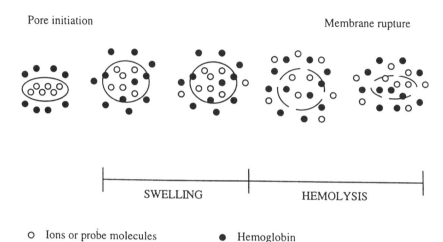

Pore initiation   Membrane rupture

SWELLING   HEMOLYSIS

o Ions or probe molecules   • Hemoglobin

FIGURE 18 Electroporation of erythrocytes based on colloid osmotic swelling. (From Ref. 15.)

Hydrophobic pore   Hydrophilic pore

FIGURE 19 Hydrophobic and hydrophilic pores. Thermal motions in the lipid molecules causes the formation of hydrophobic pores. Formation of hydrophilic pores from hydrophobic pores is called pore inversion. (From Ref. 28.)

r is small, formation of the hydrophobic pore is energetically more favorable than the formation of hydrophilic pore because a hydrophobic pore with small radius possesses less energy compared with a hydrophilic pore with equivalent radius. When the radius of the pore exceeds a critical value r^* varying between 0.3 and 0.5 nm, the pore energies of the hydrophobic and hydrophilic pores become equal. When r is greater than r^*, the energy of the hydrophilic pore is less than the energy of the hydrophobic pore. Therefore, a reorientation of the hydrophobic pores to hydrophilic pores becomes energetically favorable when r is greater than $r*$. The reorientation of hydrophobic pores to form hydrophilic pores is called the inversion of pores. The formation of hydrophilic pores with an effective radius of 0.6 to 1.0 nm increases the permeability of the membrane to ions and is known as the reversible electrical breakdown of membranes. When the radius of the hydrophilic pore exceeds r_d, the pore will grow indefinitely and result in the mechanical breakdown of the membrane.

The structure of the electric field–induced pores is illusive, and the initial formation of pores in the lipid or protein domains of the membranes is not known. The formation of pores in the lipid domain may be caused by the electric dipoles of the lipid molecule and the small but finite permeability of the lipid bilayer to ions. Electric fields cause the lipid molecules to reorient, thus creating pores and increasing the permeability of the bilayer to ions. Electric fields may also cause the generation of a electric current through the lipid layer because of the permeability of the bilayer to ions. The generation of electric currents causes local Joule heating and thermal phase transition of the lipid bilayer. The conductance of ions through the lipid bilayer occurs mainly at the lattice defects. However, the lattice defects are not static structures. Lattice defects are constantly fluctuating in the time scale of picosecond to minutes (35). Therefore, the pores formed in the lipid layer exhibit a short lifetime.

The electroporation of human erythrocyte membranes occurring in the lipid domain may be represented by Scheme 1.

$$A \rightarrow B \rightarrow C \rightarrow C' \rightarrow B' \rightarrow A \qquad \text{Scheme 1}$$

A to B represents the electroporation; B to C is the pore expansion with dipole reorientation of lipid molecules; C to C' is the formation of hydrophilic pores; and, C' to B' to A represents relaxation when the electric field is turned off (35).

Electroporation occurring in the protein channels, transport systems, or pumps may be represented by Scheme 2.

$$P_{close} \rightarrow P_{open} \rightarrow P_{denatur} \rightarrow P_{irrev} \qquad \text{Scheme 2}$$
$$\searrow [O]$$

P_{close} and P_{open} represent a membrane channel or pore in the closed and open conformations, respectively. $P_{denatur}$ represents the denatured state and P_{irrev} and [O] are the excited states of the channel or pore. In principle, the P_{close} to

P_{open} transition occurs when the membrane potential reaches the gating voltage of the channel (35).

The opening/closing of many protein channels depends on the transmembrane electric potential. The gating voltage of the protein channels (~50 mV range) is considerably smaller than the dielectric strength of the lipid bilayer. Thus, many voltage-sensitive protein channels will open to become pores before the transmembrane potential reaches the breakdown potential of the lipid bilayer. Protein channels in the open state conduct much larger current than in the close state. Due to the large current, the proteins may be irreversibly denatured by Joule heating or electrical modification of functional groups. The generation of transmembrane potential may alter the charge distribution of specific proteins, either by electroconformational change or by changes in the phosphorylation of amino acids. Redox reactions such as oxidation of -SH groups or methionine also occur in the presence of pulsed electric fields (35).

In a low ionic strength medium containing no K^+ ions , the membrane conductance generated by the externally applied electric field can be partially blocked by ouabin. Ouabin blocks the active transport of Na^+ and K^+ ions across the erythrocyte membranes. Based on this partial blockage effect of ouabin, it is reported that at least 35% of the pores are related to the opening of Na^+/K^+ ATPase channels in a low ionic strength medium. In a high ionic strength medium containing K^+ ions, ouabin does not decrease the electric field–induced membrane conductance. A large fraction of the pulsed electric field–induced pores occur at as yet unidentified sites (36).

Chernomordik et al. (25) observed a similarity in the response of artificially prepared lipid bilayers and cell membranes when subjected to electric fields. Therefore, it was suggested that the mechanism of electroporation is similar in lipid bilayers and cell membranes and that the pores developed during the electrical breakdown of cell membranes are formed in the lipid matrix. However, pores in red blood cells in a low ionic strength medium (< 30–45 mM NaCl) are formed in the channel proteins with an electric field intensity of 3.7 kV/cm, pulse duration 30 μs, and temperature 25°C (36).

VI. STRUCTURAL CHANGES IN BACTERIA AND ERYTHROCYTES

The bacterial cell membrane performs a variety of functions for the growth and survival of the cell. The membrane (37):

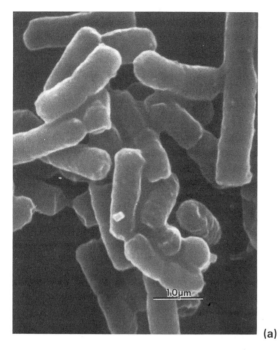

(a)

FIGURE 20   Untreated (a) and electric field treated (b) cells of *Escherichia coli* in SMUF as seen with scanning electron microscope. (From Ref. 38.)

Controls the passage of small molecules into and out of the cell.
Extrudes extracellular enzymes and cell wall materials.
Is the site of many complex activities, including RNA, protein, and
    cell wall synthesis, electron transport, and oxidative phosphoryla-
    tion.
Plays an important role in the control of DNA synthesis.

Therefore, any damage to the membrane will affect its functions and thereby
inhibit cell reproduction.

The mechanism of bacterial cell inactivation treated with pulsed electric fields was studied by Hamilton and Sale (4). Cell inactivation was caused
by plasmolysis, leakage of intracellular contents, loss of the property of *E.*

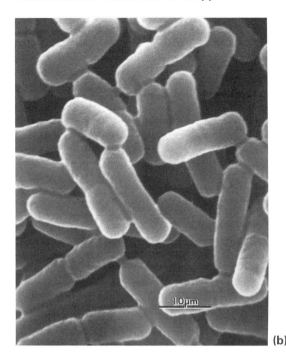

(b)

coli to plasmolyze in a hypertonic medium, and loss of β-galactosidase activity in permease-negative mutant of *E. coli*. There was a direct relationship between the number of *S. aureus* cells inactivated by pulsed electric fields and the number of cells capable of forming spheroblasts after the electric field treatment (3). Cell membranes lose the property of semipermeability when subjected to pulsed electric fields. Phase contrast microscopy indicated the cell contents shrank away from the cell wall in *E. coli* 8196 and cell plasmolysis occurred. Loss of semipermeability leads to leakage of essential molecules from the cell and finally to cell inactivation (4). No damage of cell membranes other than plasmolysis was detected using electron microscopic methods (19).

Cells of *E. coli* (ATCC 11229) suspended in SMUF were subjected to electric field treatment and observed with scanning and transmission electron microscopy. The scanning electron microscopic technique did not indicate significant differences between treated and untreated cells (Fig. 20). A micro-

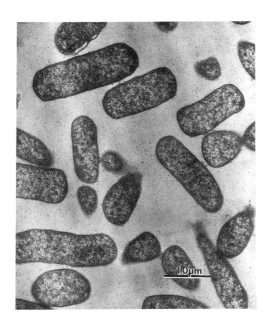

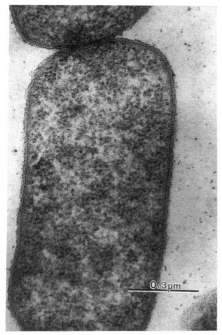

FIGURE 21 Untreated cells of *Escherichia coli* in SMUF as seen with transmission electron microscope.(From Ref. 38.)

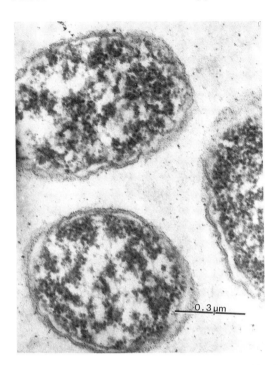

FIGURE 22   *Escherichia coli* cells suspended in SMUF and subjected to 64 pulses at 60 kV/cm and 13°C as seen with transmission electron microscope. (From Ref. 38.)

graph of thin sections of untreated control cells is presented in Figure 21. The cytoplasmic membrane and the outer membrane are close to each other. Electric field treatment results in shrinkage of the cytoplasmic membrane away from the outer membrane (Fig. 22). The crenations in the outer membrane are indicative of the shrinkage. The shrinkage is indicative of the loss of the semipermeability property of the cell membrane (38).

 S. *aureus* (ATCC 6538) suspended in SMUF and exposed to electric fields resulted in observed roughening or shrinking of the outer surface of the cell (Fig. 23). A micrograph of thin sections of untreated control cells of S. *aureus* suspended in SMUF are presented in Figure 24. The cells possess a thick cell wall typical of gram-positive bacteria. The application of 64 pulses of a 60 kV/cm electric field breaks down the cell wall in S. *aureus* (Fig. 25). The untreated cells exhibit a well-differentiated cytoplasmic orga-

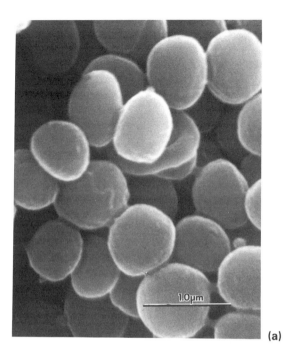

(a)

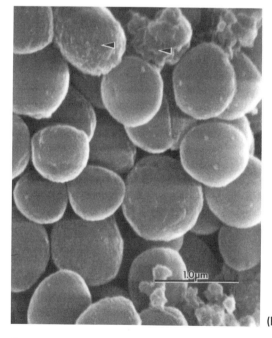

(b)

FIGURE 23   Untreated (a) and electric field treated (b) cells of *Staphylococcus aureus* in SMUF as seen with scanning electron microscope. (From Ref. 39.)

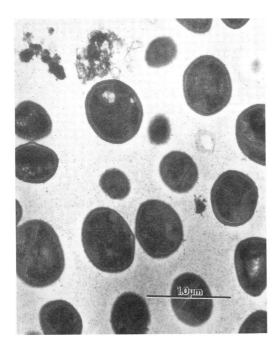

FIGURE 24 Untreated cells of *Staphylococcus aureus* in SMUF as seen with transmission electron microscope. (From Ref. 39.)

nization. However, electric field treatment disrupts the cellular organization (39).

In the case of *S. cerevisiae* (ATCC 16664), cell surface roughening, "forced budding," and elongation after PEF were seen using scanning electron microscopy (Fig. 26) (40). Figure 27 presents untreated *S. cerevisiae* cells exhibiting cellular organelles including the nuclei, vacuoles, mitochondria, and ribosomes. Most of the cellular organelles are partially or completely disintegrated after the electric field treatment. The ribosomes disintegrated into their subcomponents (Fig. 28). Electric field treatment shrinks the cytoplasmic material, suggesting leakage of the cytoplasm out of the cell. The boundary between the cell wall and cytoplasmic material is unclear at the bud scar regions in approximately 90% of the electric field treated cells, implying that only 1 out of 10 scars is formed naturally. Cytoplasmic debris is evident, most likely a result of damage to the cell walls of a few *S. cerevisiae* cells (41).

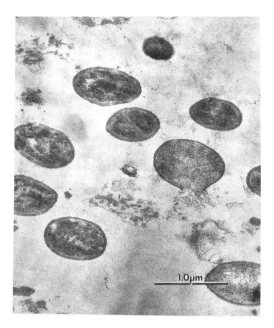

FIGURE 25 *Staphylococcus aureus* cells suspended in SMUF and subjected to 64 pulses at 60 kV/cm and 13°C as seen with transmission electron microscope. (From Ref. 39.)

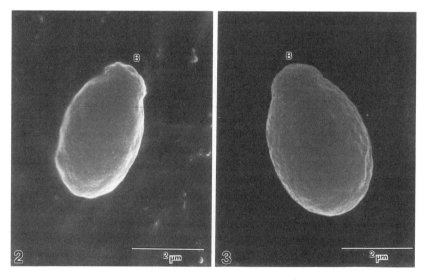

FIGURE 26 Untreated (2,3) and treated cells (4,5) of *S. cerevisiae* in apple juice as seen with scanning electron microscope. (From Ref. 40.)

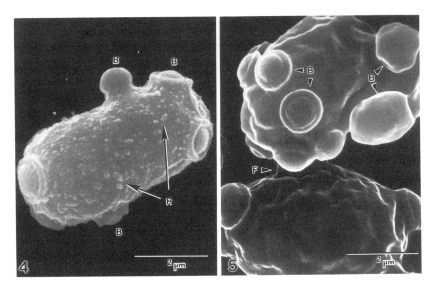

FIGURE 26   Continued

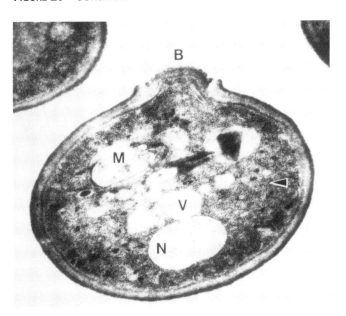

FIGURE 27   *Saccharomyces cerevisae* in untreated apple juice with indicated bud scar (B), nucleus (N), mitochondria (M), vacuole (V), and ribosomes (◄). The bar corresponds to 1 μm. (From Ref. 41.)

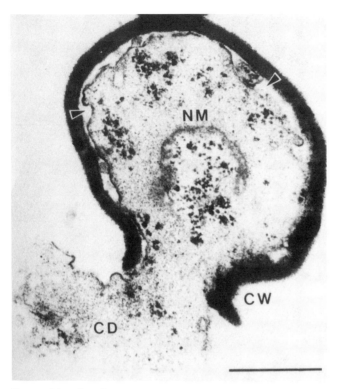

FIGURE 28   *Saccharomyces cerevisae* in electric field treated (40 kV/cm) apple juice with indicated failed nuclear membrane (NM), failed cell wall (CW), cellular debris (CD) and shrinkage of cytoplasmic material from cell wall (➤). Note the lack of cellular organelles and ribosomes. The bar corresponds to 1 μm. (From Ref. 41.)

VII.   PULSED ELECTRIC FIELDS AND ENZYMES

The activities of lipase and amylase are not inhibited by an electric field of 30 kV/cm. Cells of *E. coli* inactivated by electric fields lose the ability to synthesize β-galactosidase. However, the activity of ß-galactosidase is not affected by electric fields in *E. coli* cells incubated with benzene. No significant decrease was observed in the activities of NADH dehydrogenase (EC I.6.99.3), succinic dehydrogenase (EC I.3.99.I), and hexokinase (EC 2.7.II) after the electric field treatment (4).

Psychotrophic microorganisms cause quality problems in milk stored at refrigerated temperatures. *Pseudomonas* is a pscychrotrophic bacteria that produces protease. Proteases attack the casein and whey proteins, leading to the bitter flavor and coagulation of milk stored at 4°C. Vega et al. (42) studied the inactivation of a protease obtained from *Pseudomonas fluorescens* M3/6. The protease was inactivated up to 80% after 20 pulses, pulsing rate 0.25 Hz, at an electric field intensity of 18 kV/cm in a model system consisting of tryptic soy broth and yeast extract (Fig. 29). However, in skim milk, a 60% inactivation was obtained after 98 pulses, pulsing rate 2 Hz, with an electric field intensity of 14 kV/cm (Fig. 30). High intensity (25 kV/cm) and low pulsing

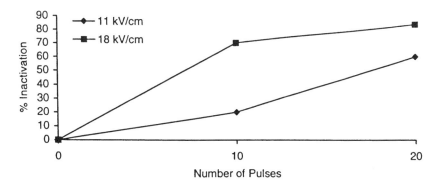

FIGURE 29 Inactivation of *Pseudomonas* M3/6 protease in a medium of tryptic soy broth with yeast extract with electric fields of 11 and 18 kV/cm and pulsing rate set to 0.25 Hz. (From Ref. 42.)

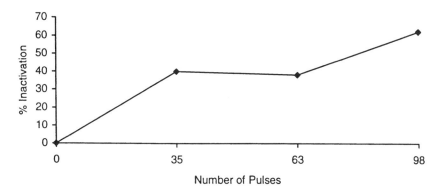

FIGURE 30 Inactivation of *Pseudomonas* M3/6 protease in a medium of skim milk with electric field of 15 kV/cm and pulsing rate set to 2 Hz. (From Ref. 42.)

rate (0.6 Hz) caused an increased susceptibility of the skim milk proteins to proteolysis.

Plasmin or milk alkaline protease is an indigenous enzyme in bovine milk. Plasmin activity results in the production of γ-casein and proteose-peptone from β-casein. The proteolytic activity of plasmin promotes a number of changes in milk, such as the viscosity decrease of casein dispersions prepared from milk, increase in the concentration of soluble proteins, increase in rennet coagulation time, increase in β-casein hydrolysis during cheese ripening and gelation of ultra-high temperature (UHT) processed milk. The activity of plasmin is reduced by thermal pasteurization, but increases again during storage. A plasmin solution was dissolved in SMUF to obtain a final concentration of 100 μg plasmin/mL SMUF. The enzyme activity was decreased by 90% after 50 pulses with an electric field intensity of 30 kV/cm at 10°C (Fig. 31). Inactivation of plasmin by thermal treatment requires 5 min at 60° or 80°C or 15 min at 40°C (43).

The enzyme alkaline phosphatase is used as an indicator of the adequacy of thermal pasteurization. The activity of alkaline phosphatase was reduced by 60% in raw and 2%-milk and by 65% in nonfat milk after 70 pulses

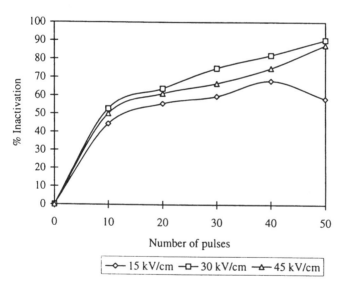

FIGURE 31   Electric field induced inactivation of plasmin (E.C.3.4.21.7.) in simulated milk ultrafiltrate. (From Ref. 43.)

at an electric field intensity of 18.8 kV/cm. Although the native alkaline phosphatase is resistant to trypsin digestion, electric field treatment rendered the alkaline phosphatase digestible to trypsin (44).

VIII. POTENTIAL IN FOOD PRESERVATION

A great deal of research is being conducted on the inactivation of microorganisms inoculated in model food systems or naturally present in foods. Electric fields can also partially inactivate specific enzymes. Significant shelf-life extensions and minimum changes in the physical and chemical properties of certain foods are demonstrable. The sensory properties of foods are not degraded by pulsed electric fields. In fact, electric field–treated green pea soup and fluid eggs are preferred over at least one of the commercial brand products. It is presumed that electric field treatment of food is accompanied by an increase in the temperature of the food, yet the maximum temperature of the food remains below thermal processing temperatures. Thus, the quality degradation associated with high temperature processing is absent or minimized in electric field treatment.

Electric field treatment is associated with minimum energy utilization and greater energy efficiency than thermal processes. In the treatment of apple juice, energy utilized in pulsed electric field technology is 90% less than the amount of energy used in the high-temperature, short-time processing method (45).

To date, much of literature discusses the application of pulsed electric fields to pumpable foods. Very little information is available on liquid foods with particulates or on solid foods. This is undoubtedly because of the difficulty in designing treatment chambers with a uniform electric field distribution (46).

There is also little information on the inactivation of microbial spores. The effects of electric field on the activity of enzymes such as polyphenol oxidase in apple juice can be studied. Although there are promising results on the inactivation of microorganisms and select enzymes, it is not wise to think of pulsed electric field technology as a separate food process. The use of other preservation methods, in combination with pulsed electric field technology, should be explored. Research combining select processing methods is under way to enhance inactivation of spores. Use of lysozyme in combination with a heat shock and pulsed electric field technology enhances inactivation of *B. subtilis* spores (17).

REFERENCES

1. B. E. Getchell, Electric pasteurization of milk. *Agric. Eng.* 16: 408-410 (1935).
2. A. J. H. Sale and W. A. Hamilton, Effects of high electric fields on microorganisms I. Killing of bacteria and yeasts. *Biochim. Biophys. Acta* 148: 781-788 (1967).
3. A. J. Castro, G. V. Barbosa-Cánovas, and B. G. Swanson, Microbial inactivation of foods by pulsed electric fields. *J. Food Proc. Pres.* 17: 47-73 (1993).
4. W. A. Hamilton and A. J. H. Sale, Effects of high electric fields on microorganisms II. Mechanism of action of the lethal effect. *Biochim. Biophys. Acta* 148: 789-800 (1967).
5. H. Hülsheger and E. G. Niemann, Lethal effects of high-voltage pulses on *E. coli* K12. *Radiat. Environ. Biophys.* 18: 281-288 (1980).
6. U. R. Pothakamury, A. Monsalve-González, G. V. Barbosa-Cánovas, and B. G. Swanson, Inactivation of *Escherichia coli* and *Staphylococcus aureus* in model food systems by pulsed electric field technology. *Food Res. Int.* 28(2): 167-171 (1995).
7. U. R. Pothakamury, A. Monsalve-González, G. V. Barbosa-Cánovas, and B. G. Swanson, High voltage pulsed electric field inactivation of *Bacillus subtilis* and *Lactobacillus delbrueckii. Revista Española de Ciencia y Tecnología de alimentos* 35(1): 101-107 (1995).
8. S. Jayaram, G. S. P. Castle, and A. Margaritis, Kinetics of sterilization of *Lactobacillus brevis* by the application of high voltage pulses. *Biotech. Bioeng* 40(11): 1412-1420 (1992).
9. Q. Zhang, A. Monsalve-González, G. V. Barbosa-Cánovas, and B. G. Swanson, Inactivation of *E. coli* and *S. cerevisiae* by pulsed electric fields under controlled temperature conditions. *Trans. ASAE* 37(2): 581-587 (1994).
10. Q. Zhang, B. L. Qin, G. V. Barbosa-Cánovas, and B. G. Swanson, Inactivation of *E. coli* for food pasteurization by high-intensity-short duration pulsed electric fields. *J. Food Proc. Pres.* 19: 103-118 (1995).
11. Q. Zhang, F. J. Chang, G. V. Barbosa-Cánovas, and B. G. Swanson, Inactivation of microorganisms in a semisolid food using high voltage pulsed electric fields. *Lebensm.-Wiss. u.-Technol.* 27(6):538-543 (1994).
12. M. V. Simpson, G. V. Barbosa-Cánovas, and B. G. Swanson, Pulsed electric field processing and the chemical composition of apple juice. Washington State University internal report.
13. J. E. Dunn and J. S. Pearlman, Methods and apparatus for extending the shelf life of fluid food products. *US Patent* 4,695,472 (1987).
14. B. L. Qin, U. R. Pothakamury, G. V. Barbosa-Cánovas, and B. G. Swanson, Food pasteurization using high intensity pulsed electric fields. *Food Tech.* 49 (12): 55-60 (1995).
15. T. Y. Tsong, Review: On electroporation of cell membranes and some related phenomena, *Bioelectrochem. Bioenergetics* 24: 271-295 (1990).

16. D. Knorr, M. Geulen, T. Grahl, and W. Sitzmann, Food application of high electric field pulses. *Trends Food Sci. Tech.* 5(3): 71-75 (1994).

17. M. V. Simpson, G. V. Barbosa-Cánovas, and B. G. Swanson, Combined inhibitory effect of lysozyme and high voltage pulsed electric fields on the growth of *Bacillus subtilis* spores. *Annual IFT meeting*, session 89, paper 2 (1995).

18. B. L. Qin, U. R. Pothakamury, G. V. Barbosa-Cánovas, and B. G. Swanson, Nonthermal pasteurization of liquid foods using high intensity pulsed electric fields. *Crit. Rev. Food Sci. Nutr.* 36 (6): 603-627 (1996).

19. H. Hülsheger, J. Potel, and E. G. Niemann, Killing of bacteria with electric pulses of high field strength. *Radiat. Environ. Biophys.* 20: 53-65 (1981).

20. Q. Zhang, A. Monsalve-González, B. L. Qin, G. V. Barbosa-Cánovas and B. G. Swanson, Inactivation of *Saccharomyces cerevisae* by square wave and exponential-decay pulsed electric field. *J. Food Process Eng.* 17: 469-478 (1994).

21. B. L. Qin, Q. Zhang, G. V. Barbosa-Cánovas, B. G. Swanson, and P. D. Pedrow, Inactivation of microorganisms by pulsed electric fields with different voltage waveforms. *IEEE Trans. Dielectrics and Electrical Insulation* 1(6): 1047-1057 (1994).

22. H. Hülsheger, J. Potel, and E. G. Niemann, Electric field effects on bacteria and yeast cells. *Radiat. Environ. Biophys.* 22: 149-162 (1983).

23. U. R. Pothakamury, H. Vega, Q. Zhang, G. V. Barbosa-Cánovas, and B. G. Swanson, Effect of growth stage and temperature on inactivation of *E. coli* by pulsed electric fields. *J. Food Prot.* 59: 1167-1171.

24. H. Vega, U. R. Pothakamury, F. J. Chang, Q. Zhang, G. V. Barbosa-Cánovas and B. G. Swanson, High voltage pulsed electric fields, pH, ionic strength and the inactivation of *E. coli. Food Res. Int.* 29 (2): 117-121 (1996).

25. L. V. Chernomordik, S. I. Sukharev, S. V. Popov, V. F. Pastushenko, A. V. Sokirko, I. G. Abidor, and Yu. A. Chizmadzhev, The electrical breakdown of cell and lipid membranes: the similarity of phenomenologies. *Biochim Biophys. Acta* 902: 360-373 (1987).

26. D. S. Dimitrov, Electric field-induced breakdown of lipid bilayers and cell membranes: a thin viscoelastic model. *J. Membr. Biol.* 78: 53-60 (1984).

27. H. G. L. Coster and U. Zimmermann, The mechanism of electrical breakdown in the membranes of *Valonia utricularis. J. Membr. Biol.* 22: 73-90 (1975).

28. R. W. Glaser, S. L. Leikin, L. V. Chernomordik, V. F. Pastushenko, and A. I. Sokirko, Reversible electrical breakdown of lipid bilayers: formation and evolution of pores. *Biochim. Biophys. Acta* 940: 275-287 (1988).

29. U. Zimmermann, G. Pilwat, F. Beckers, F. Riemann, Effects of external electrical fields on cell membranes. *Bioelectrochem. Bioenergetics* 3: 58-83 (1976).

30. L. V. Chernomordik, S. I. Sukharev, I. G. Abidor, and Yu. A. Chizmadzhev, Breakdown of lipid bilayer membranes in an electric field. *Biochim. Biophys Acta* 736: 203-213 (1983).

31. J. Teissie and T. Y. Tsong, Electric field induced transient pores in phospholipid bilayer vesicles. *Biochemistry,* 20: 1548-1554 (1981).

32. U. Zimmermann, Electrical breakdown, electropermeabilization and electrofusion. *Rev. Phys. Biochem. Pharmacol.* 105: 176-256 (1986).
33. E. H. Serpersu, K. Kinosita, and T. Y. Tsong, Reversible and irreversible modification of erythrocyte membrane permeability by electric field. *Biochim Biophys. Acta* 812: 779-785 (1985).
34. K. Kinosita and T. Y. Tsong, Hemolysis of erythrocytes by a transient electric field. *Proc. Natl. Acad. Sci. USA* 74 (5): 1923-1927 (1977).
35. T. Y. Tsong, Mini review: Electroporation of cell membranes. *Biophys. J.* 60: 297-306 (1991).
36. J. Teissie and T. Y. Tsong, Evidence of voltage induced channel opening in Na, K-ATPase of human erythrocytes membranes. *J. Membr. Biol.* 55: 133-140 (1980).
37. H. J. Rogers, H. R. Perkins, and J. B. Ward, *Microbial Cell Walls and Membranes*, Chapman and Hall, London, 1980, pp. 72-75.
38. U. R. Pothakamury, G. V. Barbosa-Cánovas, B. G. Swanson, and K. D. Spence, Effects of pulsed electric fields on the ultrastructure of *Escherichia coli*. Ph. D. thesis, Washington State Universit (1995)y.
39. U. R. Pothakamury, G. V. Barbosa-Cánovas, B. G. Swanson, and K. D. Spence, Ultrastructural changes in *Staphylococcus aureus* treated with pulsed electric fields. Ph.D. thesis, Washington State University (1995).
40. S. L. Harrison, B. G. Swanson, and G. V. Barbosa-Cánovas, High voltage pulsed electric field induced structural changes and inactivation of *Saccharomyces cerevisae*. Ph.D. thesis, Washington State University (1996).
41. S. L. Harrison, B. G. Swanson, and G. V. Barbosa-Cánovas, *Saccharomyces cerevisiae* structural changes induced by pulsed electric field treatment. *Lebensm.-Wiss. u.-Technol.* 30: 236-240 (1997).
42. H. Vega-Mercado, J. R. Powers, O. Martín Belloso, L. Luedecke, G. V. Barbosa-Cánovas, and B. G. Swanson, Inactivation of a protease from *Pseudomonas fluorescens* M3/6 using pulsed electric fields. Ph.D. thesis, Washington State University (1996).
43. H. Vega-Mercado, J. R. Powers, G. V. Barbosa-Cánovas, and B. G. Swanson, Inactivation of plasmin using high voltage pulsed electric fields. Ph.D. thesis, Washington State University (1996).
44. A. J. Castro, B. G. Swanson, G. V. Barbosa-Cánovas, and R. Meyer, Pulsed electric field modification of milk alkaline phosphatase activity. Ph.D. thesis, Washington State University (1994).
45. B. L. Qin, F. J. Chang, G. V. Barbosa-Cánovas, and B. G. Swanson, Nonthermal inactivation of *Saccharomyces cerevisae* in apple juice using pulsed electric fields. *Lebensm.-Wiss. u.-Technol.* 28: 564-568 (1995).
46. B. Qin, Q. Zhang, G. V. Barbosa-Cánovas, B. G. Swanson, and P. D. Pedrow, Pulsed electric field treatment chamber design for liquid food pasteurization using finite element method. *Trans. ASAE* 38 (2): 557-565 (1995).

5

Oscillating Magnetic Fields for Food Processing

I. INTRODUCTION

Preservation of perishable food is a very challenging subject. Introduction of processes to preserve food began near the end of the seventeenth century when Thomas Porter and John White were granted a patent for "Preserving of all kinds of foods" [1]. The food we eat is generally derived from plants or animals. Spoilage of food may result from growth of microorganisms and/or endogenous biochemical reactions. Microbial spoilage is often more pronounced and more hazardous than biochemical deterioration. Therefore, inhibition of microbial growth is important to the successful and acceptable preservation of food.

Food that is nutritious for humans is susceptible to spoilage by many microorganisms. In general, foods can be preserved by any process that inhibits growth and/or inactivates microorganisms and prevents subsequent contamination of food. Processes that involve heating foods, such as canning, inactivate microorganisms while altering quality of the food. Most other processes, such as salting, smoking, drying, pickling, and adding vinegar, sugar, and/or other chemicals, create an environment within the food that inhibits the growth of microorganisms. The utilization of oscillating magnetic fields to inactivate microorganisms has the potential to pasteurize food with an improvement in the quality and shelf life compared to conventional pasteurization processes.

This chapter reviews the effects of magnetic fields on microorganisms, cell membranes, and malignant cells and the potential applications of magnetic fields for food preservation. Magnetic fields, in general, influence the direction of migration and alter the growth and reproduction of microorganisms. Magnetic fields increase DNA synthesis, change orientation of biomolecules and biomembranes to a direction parallel or perpendicular to the applied magnetic field, and change ionic drift across the plasma membrane resulting in an altered rate of cell reproduction. Malignant cell population is reduced when placed in oscillating magnetic fields, facilitating the treatment of cancer.

II. MAGNETIC FIELDS

The region in which a magnetic body is capable of magnetizing the particles around is called the magnetic field. When the susceptibility to magnetization is equal along the three orthogonal axes x, y, and z, the particle is said to possess isotropic susceptibility. On the other hand, when the susceptibility to magnetization is unequal along the x, y, and z axes, the particle is said to possess anisotropic susceptibility. Carbon atoms exhibit isotropic susceptibility, whereas two carbon atoms bonded by single, double, or triple bonds exhibit anisotropic susceptibility (2).

Magnetic field is measured in terms of magnetic intensity B. Table 1 lists the common terms and units used to describe magnetic fields (3). Diamagnetism is the phenomenon in which the intensity of magnetization in the particle induced by an applied magnetic field is less than the magnetization induced in a vacuum. Most organic and inorganic compounds, except compounds of transition elements, are diamagnetic. Two carbon atoms bonded by single, double, or triple bonds exhibit diamagnetic anisotropy. On the other hand, paramagnetism is the phenomenon in which the intensity of the induced magnetization in a particle is more than the magnetization induced in a vacuum. Free radicals and compounds of transition elements exhibit paramagnetism (2).

In general, magnetic fields are differentiated as static (SMF) or oscillating (OMF) fields. Magnetic fields may be homogeneous or heterogeneous. In a homogeneous magnetic field, the field intensity B is uniform in the area enclosed by the magnetic coil, whereas in a heterogeneous field, B is not uniform: the intensities decrease as distance from the center of the coil increases. A heterogeneous magnetic field exerts an accelerating force on the diamagnetic and paramagnetic particles in the field, whereas a homogeneous field does not exert an accelerating force. SMF exhibits a con-

TABLE 1   Units of Magnetism

Term	CGS unit	SI unit	Relationship
Flux [a]	Line	Weber (Wb)	1 Wb = 10^8 lines
Flux density[b]			
(Wb/m^2)	Gauss (G)	Tesla (T)	1 T = 1 Wb/ m^2 = 10^4 G
Field intensity[c]	Oersted (Oe)	Ampere/meter (A/m)	1 Oe = 79.58 A/ m

CGS, centimeter-gram-second system; SI, Système International
[a]Magnetic flux is defined in terms of lines of force of a magnet. A line of force is the free path that would be traced by a unit pole in a magnetic field due to the magnetic forces acting on it.
[b]Magnetic flux density is defined as the number of lines of force per unit area. Flux density depends on the permeability of the material μ in which the field is induced: Flux density = μ* field intensity. Gauss is a unit commonly used in the industry and by manufacturers. Tesla is used more in academic and research areas.
[c]Magnetic field intensity is the force experienced by a unit pole placed in vacuum in a magnetic field. Oersted represents the unit magnetic intensity at a point in the magnetic field where the unit pole experiences a force of 1 dyne.
Source: From Ref. 3.

stant B with time, and the direction of the field remains the same. OMF, applied in the form of pulses, reverses the charge for each pulse and the intensity of each pulse decreases with time to about 10% of the initial intensity.

A. Generation of High Intensity Magnetic Fields

Magnetic fields are usually generated by supplying current to electric coils. The inactivation of microorganisms requires magnetic flux densities of 5 to 50 tesla (T). OMFs of this density can be generated by using (a) superconducting coils, (b) coils that produce DC fields, or (c) coils energized by the discharge of energy stored in a capacitor (4).

Magnetic fields with intensities up to 3 T can be generated by inserting an iron core in the coil. Above 3 T, insertion of iron core is not useful because of the magnetic saturation in the iron core; therefore, air-core solenoids are used to obtain high magnetic field intensities. The magnetic field generated by an air-core solenoid depends on the magnitude of current supplied. One of the difficulties in generating high intensity magnetic fields is the large power consumption and the Joule heat produced by the large current. Superconducting magnets can generate high intensity magnetic fields without any Joule heating. However, the superconducting magnets generate optimum magnetic fields of 20 T (5).

Magnetic fields with a flux density greater than 15 to 30 T can be obtained by using a hybrid magnet, which is a combination of a superconducting magnetic coil with a water-cooled magnetic coil insert. The superconducting coil is housed in a helium environment to provide the necessary cooling to the coil. The current required for the hybrid magnet system is ~ 40 kA (6). Magnetic fields above 30 T can be generated in a pulsed form for a short duration. The current supplied to the coil is obtained from a condenser bank.

The Magneform 7000 series™ coil (Maxwell Laboratory, San Diego, Calif.) is one such instrument that uses the energy stored in a capacitor bank. The capacitor is charged from a voltage source; when the switch is closed, completing the circuit that includes the capacitor and the coil, an oscillating current is generated between the plates of the capacitor (Fig. 1). The oscillating current generates an OMF. The frequency of the magnetic field is determined by the capacitance of the capacitor and the resistance and inductance of the coil. As the current changes direction, the magnetic field changes polarity. The oscillating current and, hence, the oscillating magnetic field rapidly

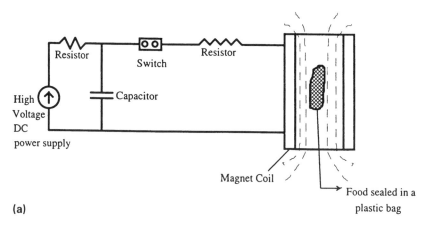

(a)

FIGURE 1 (a) Circuit for Magneform 7000 series coil, (b) Washington State University Oscillating Magnetic Field Laboratory scale unit, (c) Washington State University Oscillating Magnetic Field coil and three types of field shapers.

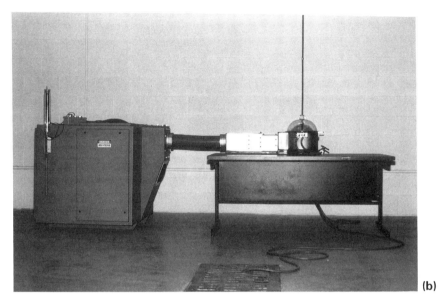

(b)

(c)

deteriorate, with the field intensity dropping to a small percentage of the original intensity after about 10 oscillations. The Magneform 7000 series™ coil was used by Hofmann (7) for studies of inactivation of food-spoilage microorganisms. The coil connected to a capacitor generates an OMF (2–50 T) in the region bounded by the coil.

For generating a magnetic field of 70 T, a capacitor with stored energy of 1.25 MJ is required (8). A multilayer, wire-wound magnet fabricated with a matrix microcomposite Cu/Nb conductor can generate a field of 68.4±1 T using a 100 kJ, 4 kV capacitor at 77°K (9). A wire made of Cu and Cr wound on a FRP bobbin and impregnated with epoxy resin can generate a field of 45 T at 77°K using capacitor banks of 200 kJ and 112 kJ (10).

Inhibition or stimulation of the growth of microorganisms exposed to magnetic fields may be a result of the magnetic field or the induced electric field. The induced electric field is measured in terms of induced electric field strength and induced current density. The induced peak electric field strength E_p is given by:

$$E_p = \frac{J_p}{\sigma} \tag{1}$$

where J_p is the peak current density given by:

$$J_p = \sigma \pi B_p r \tag{2}$$

where σ is the conductivity of the culture medium, r is the radius of the cylinder, and B_p is the peak magnetic field intensity. To differentiate between the magnetic field and electric field effects, a cylindrical enclosure containing cells and medium that can be adapted to in vitro studies employing uniform, single-phase, extremely low frequency (ELF) magnetic fields is recommended. The experimental arrangement (Fig. 2) consists of one or more cylinders, oriented with the longitudinal axis parallel to the applied magnetic field. The magnetic field is essentially uniform throughout the volume of the cylinder. The density of the induced current is zero along the axis of the cylinder and increases with an increase in distance from the center. Thus, cells placed in different circular regions of the cylinder are exposed to uniform magnetic fields of different current density. The observation of similar biological effects throughout the cylinder suggests a mechanism involving magnetic interactions with the cells, whereas an effect that differs between sections of the cylinder may be indicative of a mechanism involving the induced electric field or a combination of electric and magnetic fields (11).

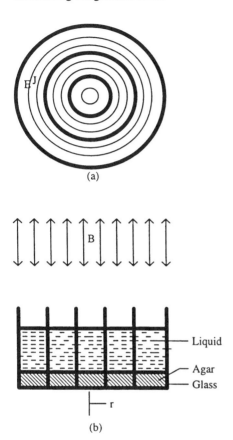

FIGURE 2 Geometry of a uniform, magnetic-field cell-enclosure combination. (a), Top view with directions of E and J schematically indicated; (b), cross-sectional view of enclosure containing agar and liquid medium. (From Ref. 11.)

III. MAGNETIC FIELDS AND MICROORGANISMS

The influence of static and/or oscillating magnetic fields on living organisms became evident in the early twentieth century with the observation of protoplasmic streaming in cells. The rate of protoplasmic streaming in algae is accelerated or retarded depending on the direction of the applied magnetic fields (12). Magnetic fields of extremely low frequency, such as the earth's magnetic field, give direction to the migration of microorganisms.

For example, certain aquatic bacteria known as magnetotactic bacteria tend to move along the lines of magnetic field and are accordingly influenced by the earth's magnetic field. By convention, the direction of a magnetic field is the one indicated by the north-seeking end of a magnetic compass needle. Accordingly, the earth's magnetic field points up at the South Pole, down at the North Pole, and horizontally northward at the equator. In the Northern Hemisphere, the magnetic field is inclined downward; in the Southern Hemisphere, it is inclined upward. In the absence of any other magnetic field except the geomagnetic field, the magnetotactic bacteria migrate northward. Therefore, in the Northern Hemisphere the north-seeking bacteria migrate downward and south-seeking bacteria migrate upward, whereas in the Southern Hemisphere the north-seeking bacteria migrate upward and south-seeking bacteria migrate downward. The influence of magnetic fields on the migration of magnetotactic bacteria is caused by the presence of magnetic particles called magnetosomes in the bacteria. The magnetosomes, essentially composed of magnetite, are synthesized by the bacteria from iron in the environment. Magnetotactic bacteria are genetically capable of synthesizing magnetosomes from available iron (13).

Adamkiewicz and Pilon (14) studied the accumulation of polysaccharide in *Streptococcus mutans* in relation to the orientation of magnetic field. Eighteen beakers containing slides dipped in a suspension of *S. mutans* were placed in a circle at equal angles. Four pairs of microscopic slides were placed in each beaker. Each pair of slides was stuck back to back with a drop of water. In each pair, one slide was labeled A and the other slide B. The slides were placed in the beaker so that the broader sides of the slides were perpendicular to the radius of the circle. In each beaker, slide A faced the outside of the circle and slide B faced the inside of the circle (Fig. 3). The accumulation of polysaccharide is 66% greater in slides facing the geomagnetic north. Polysaccharide accumulation was equal on all slides when the broad sides of the slides faced neither north nor south. In the presence of an applied magnetic field intensity ranging between 0.084 and 0.17 mT, polysaccharide accumulation progressively increases with increase in magnetic field intensity on slides facing the north. Polysaccharide accumulation on slides facing south remains constant with increasing magnetic field intensity. Increasing the magnetic field beyond 0.17 mT did not further increase the polysaccharide accumulation on slides facing north or south (14).

Exposure of paramecia to a SMF of 0.126 tesla (T) results in a considerable reduction in the swimming velocity; the movements are complex, with frequent directional changes and circular motion. Cultures placed under the influence of the earth's magnetic field serve as controls. Paramecia in a con-

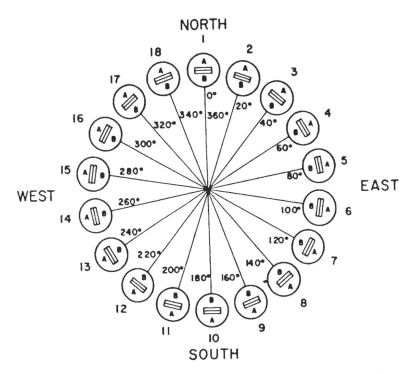

FIGURE 3 Diagrammatic representation of beakers placed in a circle for determination of polysaccharide accumulation on orientation in the geomagnetic field. (From Ref. 14.)

trol culture move in a single direction at a constant velocity. The reduction in the swimming velocity of paramecia exposed to magnetic fields is caused by the partial alignment of the diamagnetically anisotropic molecules in the cell membrane to the applied magnetic field (15).

Yoshimura (16) classified the general effects of magnetic fields on microbial growth and reproduction as (a) inhibitory effects, (b) stimulatory effects, or (c) no observable effects. Literature reports do not provide a clear understanding of the conditions under which magnetic fields have stimulatory, inhibitory, or no effects on the growth and reproduction of microorganisms.

One of the earliest reports indicates that magnetic fields do not have any effect on the morphology, growth, or reproduction of microorganisms (17). However, it was observed later that the absence of effects due to magnetic fields was caused by the homogeneity of the fields. It was also observed that a heterogeneous magnetic field inhibits budding in yeast cells. Budding was inhibited in cultures of wine yeast cells, 1- to 1.5-h-old, exposed to the magnetic field for 20, 25, 60, 120, or 150 min. There was no inhibition of budding in cultures exposed to the magnetic field for 10, 15, or 17 min. Budding was inhibited when cultures were subjected to the magnetic field for 5 min. However, the reason for this peculiar behavior was not explained (12).

Magnetic fields translocate paramagnetic free radicals such as $(C_6H_5)_3C\bullet$, $\bullet OH$, and $\bullet O$ to regions where the magnetic field intensity is the greatest, resulting in "metabolic interruptions" such as the inhibition of budding in the yeast cells. However, budding inhibition is observed only in yeast cells occupying such a position that the magnetic field is perpendicular to the "path of metabolism." Because the yeast cells are dispersed randomly in the magnetic field, budding is inhibited in only 20–30% of the cells. The rate of reproduction in yeast cells exposed for 5, 10, 20, 40, or 80 min to a homogeneous magnetic field of 1.1 T was the same as in control cultures.

Yoshimura (16) studied growth of stationary phase yeast cells subjected to SMFs. There was no change observed in the growth of yeast cells treated with an SMF having a flux density of 0.57 T. However, when the yeast cells were subjected to an OMF, the inactivation of the yeast cells increase. The reason for the increase in inactivation was not explained.

Van Nostran et al. (18) observed that the rate of reproduction of 3-day-old cells of *Saccharomyces cerevisiae* exposed to a homogeneous magnetic field of 0.46 T for 24, 48, or 72 h at 28° or 38°C decreased considerably from the rate of reproduction of control cultures. An overall interaction between temperature and magnetic field was found present only during the logarithmic phase of cell multiplication. The presence of salt increases the number of cells at 48 and 72 h. The combined effect of salt and temperature is to increase the number of cells (Table 2).

When a broth culture of *Serratia marcescens* is incubated in the presence of a heterogeneous magnetic field with an intensity of 15,000 Oersted (Oe) for 10 h at 27°C, the growth rate was similar to the growth rate of control cultures for up to 6 h. After 7 h, the culture incubated in the magnetic field contained a lower cell population than the control culture. The difference in cell population was greatest after 8 h and decreased through 10 h. After 10 h, the number of cells in the culture incubated in the magnetic field was equivalent to the control culture. Incubation of *Staphylococcus aureus* at 37°C in a magnetic field did not affect the growth rate. However, increasing the gradient of heterogeneity of the magnetic field from 2300 Oe/cm to a mean gradi-

TABLE 2   Magnetic Field Effects in the Presence and Absence of Salts

	Cells (millions)/ mL					
	Control		0% NaCl		1.5% NaCl	
Time of incubation (h)	(no magnet)		(+ magnet)		(+ magnet)	
	28°C	38°C	28°C	38°C	28°C	38°C
24	14.4	5.5	15.7	0.7	8.8	3.8
48	39	12.4	37.0	3.5	31.0	16.0
72	40.5	12.5	33.0	7.9	31.4	17.0

Source: From Ref. 18.

ent of 5200 Oe/cm resulted in a greater cell population after 3 to 6 h of incubation compared to the controls. After 6 h, growth was inhibited, and after 7 h, the cell population of *S. aureus* was lower than the cell population of control culture. Between 7 and 9 h, the growth rate increased, and the cell population of *S. aureus* at 9 h was equivalent to controls. The decrease in the growth rate was attributed to the increase in the inactivation of the cells and not to the decrease in the bacterial fission rate (19).

Moore (20) reported maximum stimulation of the growth of *Pseudomonas aeruginosa, Halobacterium halobium*, and *Candida albicans* in an OMF of 0.015 T and frequency of 0.3 Hz, and maximum inhibition in an SMF of 0.03 T. The inhibition of cell reproduction was attributed to the decrease in the rate of cell division and not to the inactivation of microorganisms, contrary to the conclusions of Gerencser et al. (19). There was no change in the morphology of the cells and no observable effect, such as change in mutation frequency, on the genetic apparatus of the cell when subjected to the magnetic fields, though changes in the genetic apparatus cannot be completely ruled out. The temperature of the culture increased by 0.1°C during exposure to the magnetic fields.

IV. MAGNETIC FIELDS, TISSUES, AND MEMBRANES

Biological membranes exhibit strong orientation in a magnetic field because of the intrinsic anisotropic structure of the membranes. Orientation of cell membranes parallel or perpendicular to the applied magnetic fields depends on the overall anisotropy of the biomolecules, such as proteins associated with the membrane. The peptide bond in the protein molecules resonating between the two structures (Fig. 4) exhibits diamagnetic anisotropy and therefore orients parallel to an external magnetic field, as does the plane of the carbon–carbon double bond (21).

Alterations in ion flux across the plasma membrane can alter cell division rates. SMFs in the range of about 100 T are required to increase or decrease the ion flux across the cell membrane (22). When an OMF with peak flux density of 5.4 mT and extremely low frequency of 72 Hz was applied to cells of paramecia for 24 h, the daily cell division rate increased by 4.7, 8.5, and 16.8%, respectively, in the Pawn mutant, wild-type, and Paranoic mutant paramecia compared to controls. Cultures grown under the influence of geomagnetic fields served as controls (23). To verify the dependence of paramecia cell division on ionic flux across the membrane, verapamil, a calcium channel blocker in cardiac tissue, was used to block the Ca^{2+} channel in the plasma membrane. The cell division rates of paramecia were equivalent in the magnetized and unmagnetized cultures when verapamil was added to the culture medium.

Dihel et al. (23) reported that the plasma membrane is a very important site of alterations when microorganisms are exposed to magnetic fields. A probe, $diSC_3(5)$ senses the changes in membrane potential by altering its partitioning across the lipid bilayer. The fluorescence of 1 μM $diSC_3(5)$, with excitation wavelength set at 622 nm and emission wavelength from 650 to 750 nm, was less in membranes exposed to magnetic fields than the fluorescence in the controls.

A change in membrane fluidity resulted in a change in the membrane-associated enzymatic activity and Ca^{2+} binding to the phospholipid head groups. Measurement of membrane fluidity demonstrated that magnetized membranes exhibited less fluidity than the controls (23).

Dystrophic changes occur in liver tissues of rats exposed to magnetic fields. The control group was under the influence of the earth's magnetic field; two experimental groups were subjected to higher magnetic field intensities. One experimental group was used for ultrastructural studies, and the other for mitochondrial respiration studies. The two experimental groups were sub-

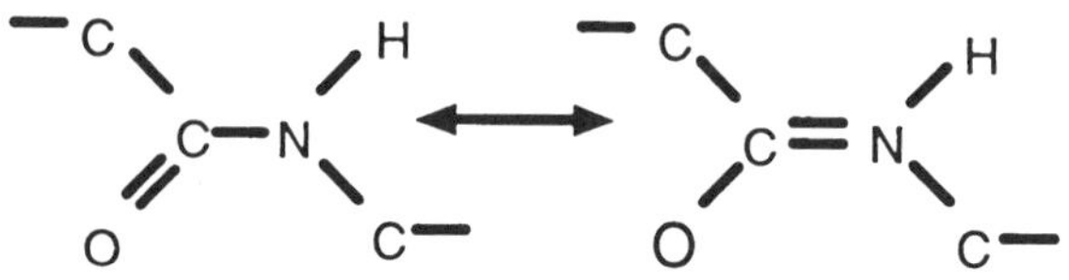

FIGURE 4   Resonating structures of peptide bond. (From Ref. 21.)

jected to an SMF of 10^{-2} T or 10^{-3} T for 1 h/day for a period of 10 days, with an interval of 23 h between the exposures (24).

Cytoplasm disintegration, irregular chromatin distribution, and an increase in the number of cytoplasmic vacuoles were observed in hepatocytes of rats after the magnetic field exposure. During the first 4 days of exposure, there were no structural changes in the hepatocytes, but after 6 days a considerable loss of endoplasmic reticulum and ribosomes was observed. Changes in the mitochondria ranging from swelling to disintegration of the matrix and reduction of mitochondrial cristae were observed after exposure. The external membranes of the hepatocytes and nuclei were not damaged. The degree of the structural change in hepatocytes was proportional to the duration of exposure to magnetic fields. The glycogen in the exposed hepatocyte cells is coarse-grained, whereas glycogen in the untreated cells is smooth-grained (24).

No change in activities of mitochondrial respiratory enzymes was observed during the first 5 days of exposure to the magnetic fields, but the activities of NADH, succinic dehydrogenases, and cytochrome oxidase in hepatocyte mitochondria increased between 6 and 8 days. Paramagnetic properties of iron-storing organs such as the liver, spleen, and bone marrow render the organs susceptible to the effects of magnetic fields. There was no increase in the body temperature of experimental rats exposed to the magnetic fields (24).

DNA synthesis in stationary phase cells of human fibroblasts increased when subjected to a magnetic field of 2.3 to 56 μT and frequencies in the range of 1.5 to 4 kHz. The threshold magnetic field intensity for the increase in DNA synthesis is 5 to 25 μT (25).

Ross (26) studied the effects of a combination of SMFs and extremely low frequency (ELF) magnetic fields on the cell growth of rabbit ligament fibroblasts. Helmholtz coils placed on the surface of an incubator shelf were used to generate the magnetic fields. The growth of cells was either inhibited or stimulated, depending on the SMF and strength and frequency of the ELF fields. Inhibition of cell growth occurred at ELF magnetic fields below 0.5 mT, and stimulation of cell growth occurred at ELF magnetic fields above 0.6 mT. The frequency of ELF field was held constant at 100 Hz and the strength of SMF was held constant at 0.13 mT. The transition from inhibition to stimulation was marked, occurring between 0.5 and 0.6 mT.

Electromagnetic field (EMF) in the range of $(1.6–3.6) \times 10^4$ ampere/m accelerates the malonic dialdehyde (MDA) accumulation during the Fe^{2+}-ascorbate induced oxidation of liposomes from phospholipids. The accumulation of MDA depends on temperature, concentration of the lipid, and intensity of the magnetic field. The accumulation of MDA is not affected by magnetic field in the temperature range of 8–17°C, but increases in the temperature range of 20–25°C, and decreases in the temperature range of 30–37°C. The rate of MDA accumulation decreases with a decrease in lipid concentration of liposomes (27).

V. MAGNETIC FIELDS AND MALIGNANT CELLS

Costa and Hofmann (28) discovered that high intensity magnetic fields of 1 to 50 T with a frequency of 5 to 1000 kHz will reduce the concentration of malignant cells within the tissue of a living animal (rat). Generally, animals or animal body parts afflicted with malignant cells are placed in magnetic field and subjected to a total of 1 to 1000 pulses depending on the type of tumor, with 10 pulses in 100 μsec to 1 sec for each therapy session. OMFs reduce the cell population below a threshold concentration where natural antitumor mechanisms are sufficient to counter the growth of the residual population of malignant cells. The tumors diminished in size relative to untreated tumors. No heat was generated in body tissues exposed to the magnetic fields. Normal tissues also undergo some damage due to the applied magnetic fields, but the damage is much less than that from irradiation of malignant cells. Furthermore, magnetic fields cause less damage to the natural immune system of the body compared with ionizing irradiation, so the repair and regeneration of healthy tissues require less time.

VI. MAGNETIC FIELDS AND ENZYMATIC REACTIONS

Rabinovitch et al. (29,30) and Maling et al. (31) studied the effects of static magnetic fields on three enzyme substrate systems:

1. Ribonuclease-RNA and cytocrome "c" reductase-succinate subjected to a magnetic field up to 4.8 T for 5 to 6 min (31).
2. Ribonuclease-RNA, horseradish peroxidase—dianisidine tyrosinase-L-tyrosine, and aldolase-FDP (fructose 1,6 diphosphate) systems exposed to a magnetic field of 8.5 to 17 T with exposure time of 2 to 20 min (29).
3. Trypsin-*N*-benzoyl DL-arginine p-nitroanilide (BAPA) was exposed to a magnetic field of 22 T for 9 min, and trypsin alone was "pretreated" for 65 to 220 min at 20.8 T before assaying enzyme activity toward the substrate (BAPA) (30).

Magnetic fields do not have an appreciable effect on the rates of enzyme-catalyzed reactions.

The activity of carboxydismutase in *tris* buffer at pH 8.0 decreased when exposed to a magnetic field of 2 T at 4°C. In another study, the same enzyme was used as control and experimental agent. The enzyme served as experimental sample when the magnetic field was on, and as a control when the magnetic field was off. Enzyme activity increased considerably when the magnet is turned on, whereas activity decreases as soon as the magnet is turned off. The activating effect of the magnetic field is caused by an increase in hydrogen bonding and, consequently, the helicity of the polypeptide backbone of the enzyme. The increase in helicity and hydrogen bonding stabilizes the protein against denaturation; thus the enzyme is less susceptible to inactivation (32).

VII. MECHANISMS OF MAGNETIC FIELD INACTIVATION OF MICROORGANISMS

A. ICR and IPR Models

a. The ion cyclotron resonance (ICR) model

An ion entering a magnetic field B at a velocity v experiences a force F given by:

$$\vec{F} = q\, (\vec{v} \times \vec{B}) \tag{3}$$

When the v and B are parallel, the force F is zero (Fig. 5). When v is normal to B, the ion moves in a circular path (Fig. 6). For other orientations between v and B, the ion moves in a helical path (Fig. 7). The frequency at which the ion revolves in the magnetic field is known as the gyrofrequency υ, which depends on the charge-to-mass ratio of the ion and the magnetic field intensity:

$$\upsilon = q\, B\, /\, (2\, \pi m) \tag{4}$$

where q is the charge and m is the mass of the ion. Cyclotron resonance occurs when υ is equal to the frequency of the magnetic field. At 50 μT, the resonance frequency of Na^+ and Ca^{2+} are 33.3 and 38.7 Hz, respectively. At cyclotron resonance, energy is transferred selectively from the magnetic field to the ions.

Energy is also transferred to the metabolic activities involving the ions.

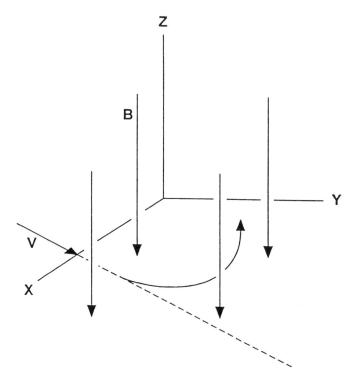

FIGURE 5   Charged particle in a magnetic field. (From Ref. 33.)

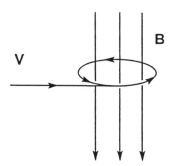

FIGURE 6   Charged particle in a magnetic field when v is normal to B. (From Ref. 33.)

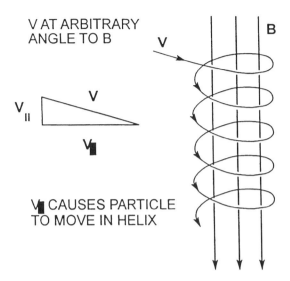

V AT ARBITRARY ANGLE TO B

V⊥ CAUSES PARTICLE TO MOVE IN HELIX

FIGURE 7   Charged particle in a magnetic field when v makes an arbitrary angle with B. (From Ref. 33.)

Transfer of energy to the ions results in an increase in the velocity and ionic drift and therefore in an increase in the net transport of ions such as Ca^{2+} across the cell membrane. An increase in the efflux of Ca^{2+} is frequency specific, whereas changes induced in the metabolic activities occur over a range of frequencies.

The interaction sites in the magnetic field are the cell tissues most affected by the magnetic field. The ions transmit the effects of magnetic fields from the interaction site to other tissues and organs. For example, Figure 8 suggests that the cell membrane is the interaction site and the effect of the magnetic fields is transmitted to the cytoskeleton, organelles, nuclear membrane, chromosome, and protein molecules. Therefore, the intensity of response to the magnetic field is diffused and delayed in the tissues other than the interaction site (Fig. 9) (33).

b. The ion parametric resonance (IPR) model

The IPR model predicts the biological results from exposure to low intensity electric and magnetic fields. The model considers the biological response to parallel AC and DC magnetic fields. The functional influences of magnetic field parameters are specified, providing detailed predictions of expected

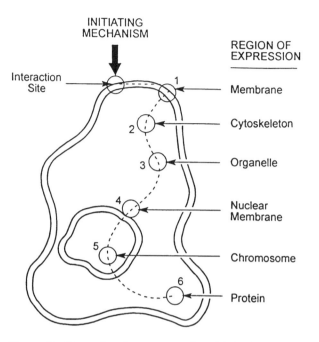

FIGURE 8   Cascade of responses in a biological cell exposed to magnetic field. (From Ref. 33.)

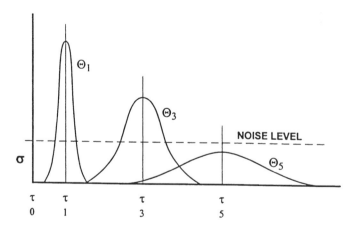

Figure 9   Intensity of response in regions 1, 3, and 5 in Figure 8. $Q_1 > Q_3 > Q_5$. (From Ref. 33.)

atomic level responses. The IPR model predicts ion response correlated with changes in experimentally controlled variables, namely magnetic flux densities of the AC and DC magnetic fields, AC frequency, and charge to mass ratio. The model is based on the concept that changes in the interactions of specific ions with biological matrices (e.g., proteins) lead to consistent observable changes at the cellular level. Biological activity is driven by enzymatically controlled chemical reactions, with some enzymes incorporating specific ions as cofactors to initiate or modulate the reaction rates. "When not exposed to AC magnetic fields, ions sitting in special crevices in cellular enzymes may vibrate isotropically with an equal probability of bouncing off every single portion of the cavity. But under certain specific conditions— when the AC field is parallel to the Earth's magnetic field and of certain intensities, AC fields may cause ions to vibrate anisotropically, altering the vibrational structure of enzyme and hence the activity" (34). The ICR model predicts the response of specific ions when the frequency of the magnetic field corresponds with the ICR conditions different for each ion. The ICR model, however, does not indicate how the ion response might vary with different magnetic flux densities (34,35).

The IPR model suggests that during exposure to magnetic fields at ion resonance frequency, the interaction of an ion with the biomolecular environment may change in a measurable and predictable way across a range of intensity values of magnetic field B. The change in interaction may be seen as a change in the biological activity. However, at "off-resonance" conditions (i.e., no biologically significant ion at resonance), the response of the biological system is unchanged across a comparable range of intensity values of B. Most of the earlier models were largely descriptive rather than predictive. The earlier models considered the interaction when the cells are oriented in a direction either in parallel or perpendicular to the applied magnetic field (35).

One of the earlier models proposed by Lednev (35) considers only the effects of ions bound in ligand structures specific to Ca^{2+}. The IPR model considers the potential effects on any unhydrated ion presumably bound within a molecular structure that can influence the observed biological response. The molecular structure may be composed of proteins, nucleic acids, or lipids, as long as the molecular structure itself requires an ionic cofactor to function (35).

Three critical tests will identify whether a biological systems response to parallel AC and DC magnetic fields is consistent with the IPR model. The tests are based on controlled variations in three basic parameters: the flux densities of parallel components of the AC and DC magnetic fields and the AC magnetic field frequency to examine on and off resonance cases. For "on-resonance" tests the response of ions is nonlinear. The "off-resonance" response is predicted to be constant across a range of AC flux densities (35).

B. Other Theories

When electrical and magnetic fields interact with a biological system, the force exerted, F, on a charged particle in the system is given by (36):

$$\vec{F} = q\,(\vec{E} + \vec{v} + \vec{B}) \qquad (5)$$

where q is the effective charge, E is the electric field, v is the velocity of the particle, and B is the magnetic field. The magnetic field effects may occur through the term (v × B). The force generates an ion current that in turn leads to the transport of ions from one location to another, such as from one side of the membrane to the other. Additionally, the forces can result in small displacements of fixed charges that may open and close voltage sensitive membrane channels. The force exerted on a particle with a dipole moment, m, and magnetic flux gradient, ∇B, is given by:

$$\vec{F} = (\vec{m} \bullet \vec{\nabla})\,\vec{B} \qquad (6)$$

Magnetic fields can apply torque on magnetic dipoles when the particles forming the dipoles are arranged in a string as in bacteria. The torque exerts a large enough force on membranes to open channels in cell membranes.

The literature provides theories to explain inactivation mechanisms for cells placed in static or oscillating magnetic fields. Two theories were proposed to explain the cell growth–inhibiting effect of "weak" OMF on living organisms. Valeri Lednev, a biophysicist at the Soviet Institute of Biological Physics in Puschino (37), proposed that a "weak" OMF can loosen the bonds between ions and proteins. British researchers John Male and Donald Edmond observed that in the presence of a steady background magnetic field such as that of the earth, the biological effects of OMF are more pronounced around the cyclotron resonance frequency of ions (37).

A second theory considers the effect of SMFs and OMFs on calcium ions bound by a calcium-binding protein such as calmodulin. The calcium ions continually vibrate about equilibrium positions in the binding sites of calmodulin. Applying a steady magnetic field to calmodulin causes the plane of vibration to rotate, or precess, in the direction of the magnetic field at one-half the cyclotron frequency of the bound calcium. Adding a "wobbling" magnetic field at the cyclotron frequency disturbs the precession and results in "loosening" of the bond between the calcium ion and the calmodulin (37).

Loosening of the bond potentially affects the metabolic activities of the biological cell.

Hofmann (7) explained the OMF inactivation of microorganisms based on the theory that the OMF may couple energy into the magnetically active parts of large critical molecules such as DNA. Within the magnetic field of 5 to 50 T, the amount of energy per oscillation coupled to one dipole in the DNA is 10^{-2} to 10^{-3} ev. With several oscillations and a collective assembly of dipoles, local activation may result in the breakdown of covalent bonds in the DNA molecule and inhibit the growth and reproduction of microorganisms.

Several theories are proposed to explain the inactivation of malignant cells. One theory suggests that the malignant cells are susceptible to inactivation by OMF because the magnetic field creates eddy currents in the tumor cells. The localized eddy currents inactivate or inhibit reproduction of malignant cells. A second theory attributes inactivation to the breaking down of magnetically susceptible macromolecules in malignant cells. Once the macromolecules are broken down, the malignant cells become inactive or no longer reproduce. Yet another theory is that the OMFs interfere with the transfer of free radicals or electrons through a chain of macromolecules unique to malignant cells (28).

VIII. MAGNETIC FIELDS IN FOOD PRESERVATION

Many methods of food preservation significantly alter the nature of the food or impart undesirable characteristics. For example, milk attains a cooked flavor when subjected to thermal processing. Foods such as cheese and beer are products of fermentation by microorganisms. Fermentation beyond a required extent results in the deterioration of food products. Therefore, the microorganisms should be inactivated after the desired fermentation. The OMF technology will be useful in inactivating the microorganisms after the desired fermentation.

Hofmann (7) observed the inactivation of microorganisms exposed to OMF with a flux density greater than 2 T. A single pulse with a flux density between 5 and 50 T and frequency of 5 to 500 kHz reduces the number of microorganisms by at least two log cycles. The technology of inactivating microorganisms by placing the microorganisms in magnetic fields may be used to improve the quality and increase the shelf life of pasteurized foods.

The most important requirement of food successfully preserved with magnetic field technology is high electrical resistivity, greater than 10 to 25 ohms-cm. Many foods have electrical resistivity in this range. For example, the electrical resistivity of orange juice is 30 ohms-cm. The applied magnetic field intensity is a function of the electrical resistivity and thickness of the food being magnetized, with larger magnetic field intensities used for smaller electrical resistivities and greater thicknesses. The food systems preserved with magnetic fields were (a) milk with *Streptococcus thermophilus*, (b) yogurt with *Saccharomyces*, (c) orange juice with *Saccharomyces*, and (d) Brown 'N Serve™ rolls dough with bacterial spores (7).

Preservation of foods with magnetic field involves sealing the food in a plastic bag, subjecting it to 1 to 100 pulses in an OMF with a frequency between 5 and 500 kHz at a temperature of 0° to 50°C for a total exposure time ranging from 25 μs to 10 ms. The exposure time is equal to the number of pulses multiplied by the duration of each pulse. The duration of each pulse includes 10 oscillations. After 10 oscillations, the substantially decayed magnetic field has negligible effect. A metal package cannot be used in a magnetic field.

No special preparation of food is required before treatment of the food by OMF. Frequencies higher than 500 kHz are less effective for microbial inactivation and tend to heat the food material. Magnetic field treatments are carried out at atmospheric pressure and at a temperature that stabilizes the food material. The food is sterilized without any detectable change in quality. The temperature of the food increases by 2° to 5°C, and the organoleptic properties change very little after magnetic field treatment. The inactivation of microorganisms in food subjected to OMF pulses is presented in Table 3.

Magnetic fields exhibit considerable bactericidal effect on water. The potential application of magnetic fields for decontamination and recycling of water on space flights was tested by Chizhov et al. (38). Magnetic fields with

TABLE 3   Magnetic Field Inactivation of Food Spoilage Microorganisms

Type of food	Temp-erature (°C)	Field intensity (Tesla)	No. of pulses	Frequency of pulses (kHz)	Initial bacterial count/mL	Final bacterial count/mL
Milk	23.0	12.0	1	6.0	25,000	970
Yogurt	4.0	40.0	10	416.0	3,500	25
Orange juice	20.0	40.0	1	416.0	25,000	6
Brown'N Serve rolls dough		7.5	1	8.5	3,000	1

Source: From Ref. 7.

intensities of 500, 1000, and 1500 Oe were used to treat water containing 10^3, 10^4, and 10^5 CFU of *E. coli* (strain no. 370) per milliliter. One-day-old culture of *E. coli* was inoculated into the water. However, the rate of inactivation of *E. coli* is lower when water contains higher concentration of *E. coli* .

Food preservation by application of magnetic fields is safe to perform. The high intensity magnetic field exists only within the coil and the immediate vicinity. Within a very short distance from the coil, the intensity of the magnetic field drops drastically. For example, if the intensity inside the coil is about 7 T, within about 2 m beyond the coil the intensity drops off to about 7 $\times 10^{-5}$ T, comparable to the geomagnetic field intensity. Thus, an operator positioned a reasonable distance from the coil is out of danger, and the magnetic field may be operated without shielding. Microwave cooking also involves application of magnetic fields. In microwave cooking, the applied magnetic field produces an induced thermal effect. Microbial inactivation in microwave systems occurs as a result of the induced thermal effects.

IX. CONCERNS ABOUT LOW INTENSITY, LOW FREQUENCY MAGNETIC FIELDS

The results of a number of studies (34) suggest that low intensity, low frequency electric and magnetic fields may influence physiological processes in biological systems. However, most theoretical models are unable to establish a predictive association between low intensity field exposure and biological results. Many experimental reports fail to document relevant field exposure parameters and do not establish a clear reproducible protocol. Inconsistencies between experimental results may be interpreted as evidence that electric and magnetic fields may not be the causal factors of the biological responses. Reproducible data can be obtained when experimental conditions are carefully replicated. Although the public in many parts of the world is pushing to set standards for 60 Hz magnetic fields, the scientific evidence is not strong enough to establish rational limits (34).

X. FUTURE RESEARCH

The inhibitory effect of SMFs or OMFs on the growth of microorganisms exhibits the potential to inactivate microorganisms in food. The technological advantages of inactivating microorganisms with OMFs include (a) minimal thermal denaturation of nutritional and organoleptic properties, (b) reduced energy requirements for adequate processing, and (c) potential treatment of foods inside a flexible film package to avoid postprocess contamination. The conditions necessary for magnetic fields to exhibit inhibitory, stimulatory, or no effects on microorganisms are not clearly understood. Though several mechanisms are proposed to explain the inhibitory effects of magnetic fields on microorganisms, there is little explanation for the stimulatory effects. Also, reported inactivation of the microbial population is limited to two log cycles. For commercialization of OMF technology, more effective and uniform inactivation of microorganisms will be required.

OMF treatment of foods in containers ensures no postprocess contamination, unlike conventional thermal processes. The thermal energy input during exposure of food to magnetic fields is small. Magnetic fields of 5 to 50 T with short pulses of 25 μs to 10 ms and frequencies of 5 to 500 kHz are required for the inactivation of microorganisms. Additional research is necessary to correlate the inactivation of microorganisms in food to the OMF flux density, OMFs to the denaturation of nutritional components of food, and the energy efficiency of magnetic fields to the increased shelf life of food. The effects of magnetic fields on the quality of food and the mechanism of inactivation of microorganisms must be studied in more detail.

REFERENCES

1. S. Thorne, *The History of Preservation of Food*, Parthenon Pub. Group Ltd., Cumbria (1986).
2. L. N. Mulay, Basic concepts related to magnetic fields and magnetic susceptibility *Biological Effects of Magnetic Fields* (M. F. Barnothy, ed.), Plenum Press, New York, 1964, Vol. 1, pp. 33-55.
3. R. A. Abler, Magnets in biological research, *Biological Effects of Magnetic Fields* (M. F. Barnothy, ed.), Plenum Press, New York, 1969, Vol. 2, pp. 1-27.
4. R. Gersdorf, F. R. de Boer, J. C. Wolfrat, F. A. Muller, L. W. Roeland, The high magnetic facility of the University of Amsterdam, *High Field Magnetism*, Proceedings of the Intl. Symp. on High Field Magnetism (M. Date, ed.), Osaka, Japan, Sept. 13-14, 1983, pp. 277-287, North Holland Pub. Co., Amsterdam.
5. N. Miura, Generation of megagauss magnetic fields and their application to solid state physics, *Physics and Engineering Applications of Magnetism* (Y. Ishikawa and N. Miura, eds.), Springer series in Solid State Sciences, 92, Springer-Verlag, Berlin, 1991, pp. 19-47.
6. L. G. Rubin, R. J. Weggel, M. J. Leupold, J. E. C. Williams, and Y. Iwasa, High field facilities at the Francis Bitter National Magnet laboratory, *High Field Magnetism*, Proceedings of the Intl. Symp. on High Field Magnetism (M. Date, ed.), Osaka, Japan, Sept. 13-14, 1983 North Holland Pub. Co., Amsterdam, pp. 249-254.

7. G. A. Hofmann, Deactivation of microorganisms by an oscillating magnetic field. U. S. Patent 4,524,079 (1985).

8. A. Yamagishi and M. Date, High magnetic field facility in Osaka University, *High Field Magnetism*, Proceedings of the Intl. Symp. on High Field Magnetism (M. Date, ed.), Osaka, Japan, Sept. 13-14, 1983, North Holland Pub. Co., Amsterdam, pp. 289-298.

9. S. Foner, 68.4 T long pulse magnet: test of high strength microcomposite Cu/Nb conductor. *Appl. Phys. Lett.* 49: 982-983 (1986).

10. N. Miura, T. Goto, K. Nakao, S. Taeyama, T. Sakakibara, and F. Herlach, Megagauss magnetic fields—Generation and application to magnetism. *J. Magnetism Magnetic Materials* 54-57: 1409-1414 (1986).

11. M. Misakian and W. T. Kaune, Optimal experimental design for in vitro studies with ELF magnetic fields. *Bioelectromagnetics* 11: 251-255 (1990).

12. G. C. Kimball, The growth of yeast in a magnetic field. *J. Bacteriol.* 35: 109-122 (1938).

13. R. P. Blakemore and R. B. Frankel, Magnetic navigation in bacteria. *Sci. Am.* 13: 58-65 (1981).

14. V. W. Adamkiewicz and D. R. Pilon, Magnetic stimulation of polysaccharide accumulation in cultures of *Streptococcus mutans. Can. J. Microbiol.* 29: 464-467 (1983).

15. M. S. Rosen and A. D. Rosen, Magnetic field influence on paramecium motility. *Life Sciences* 46: 1509-1515 (1990).

16. N. Yoshimura, Application of magnetic action for sterilization of food. *Shokuhin Kaihatsu* 24(3): 46-48 (1989).

17. M. W. Jennison, The growth of bacteria, yeasts, and molds in a strong magnetic field. *J. Bacteriol.* 33: 15-16 (1937).

18. F. E. Van Nostran, R. J. Reynolds, and H. G. Hedrick, Effects of high magnetic field at different osmotic pressures and temperatures on multiplication of *Saccharomyces cerevisiae. Appl. Microbiol.* 15: 561–563 (1967).

19. V. F. Gerencser, M. F. Barnothy, and J. M. Barnothy, Inhibition of bacterial growth by magnetic fields. *Nature* 196: 539-541 (1962).

20. R. L. Moore, Biological effects of magnetic fields: studies with microorganisms. *Can. J. Microbiol.* 25: 1145-1151 (1979).

21. G. Maret and K. Dransfield, Biomolecules and polymers in high steady magnetic fields, *Topics in Applied Physics. Strong and Ultrastrong Magnetic Fields and their Applications* (F. Herlach, ed.), Springer-Verlag, Berlin, 1985, pp. 143-204.

22 R. L. Liboff, Theoretical aspects of magnetic field interactions with biological systems *Magnetic Field Effect on Biological Systems*, Proceedings of Biomagnetic Effects Workshop (T. S. Tenforde, ed), Lawrence Berkeley Laboratory, Univ. of California April 6, 1978, Plenum Press, New York, pp. 81-82.

23. L. E. Dihel, J. Smith-Sonnenborn, and R. C. Middaugh, Effects of extremely low

frequency electromagnetic field on cell division rate and plasma membrane of *Paramecium tetraurelia. Bioelectromagnetics* 6: 61-71 (1985).

24. E. Gorczynska and R. Wegrzynowicz, Structural and functional changes in organelles of liver cells in rats exposed to magnetic fields. *Environ. Res.* 55: 188-198 (1991).

25. A. R. Liboff, T. Williams, D. M. Strong, and R. Wistair, Time varying magnetic fields: effect on DNA synthesis. *Science* 223: 818-820 (1984).

26. S. M. Ross, Combined DC and ELF magnetic fields can alter cell proliferation. *Bioelectromagnetics* 11: 27-36 (1990).

27. V. M. Aristarkhov, L. L. Klimenko, A. I. Deyev, and Y. V. Ivanekha, Influence of a constant magnetic field on the processes of peroxide oxidation of lipids in phospholipid membranes. *Biophysics* 28 (5): 848-855 (1983).

28. J. L. Costa and G. A. Hofmann, Malignancy treatment. U. S. Patent 4,665,898 (1987).

29. B. Rabinovitch, J. E. Maling, and M. Weissbluth, Enzyme substrate reactions in very high magnetic fields. I. *Biophys. J.* 7: 187-204 (1967).

30. B. Rabinovitch, J. E. Maling, and M. Weissbluth, Enzyme substrate reactions in very high magnetic fields. II. *Biophys. J.* 7: 319-327 (1967).

31. J. E. Maling, M. Weissbluth, and E. E. Jacobs, Enzyme substrate reactions in high magnetic fields. *Biophys. J.* 5: 767-776 (1965).

32. G. Akoyunglou, Effect of a magnetic field on carboxydismutase. *Nature* 202: 452-454 (1964).

33. A. R. Liboff, Cyclotron resonance in membrane transport, *Interactions between Electromagnetic Fields and Cells* (A. Chiabrera, C. Nicolinis, H. P. Schwan, eds.)Series A: Life Science 97, Plenum Press, New York, 1985, pp. 281-296.

34. B. Hileman, Findings point to complexity of health effects of electric and magnetic fields. *Chem. Eng. News* 72 (29): 27-33 (1994).

35. J. P. Blanchard and C. F. Blackman, Clarification and application of an ion parametric resonance model for magnetic field interactions with biological systems. *Bioelectromagnetics* 15: 217-238 (1994).

36. F. S. Barnes, Mechanisms for biological changes resulting from electric and magnetic fields. Proceedings of the Japan-US Science Seminar on Electromagnetic Field Effects Caused by High Voltage Systems, 1994, Paper 8-1.

37. A. Coughlan and N. Hall, How magnetic field can influence your ions? *New Scientist*, Aug. 4, p. 30 (1990).

38. S. V. Chizhov, Yu. Ye. Sinyak, M. I. Shikina, S. I. Ukhanova, and V. V. Krasnoshchekov, Effect of magnetic field on *E. coli. Moscow Kosmicheskaya Biologiya I Aviakosmicheskaya Meditsina* (in Russian) 9 (5): 26-31 (1975).

6

Application of Light Pulses in the Sterilization of Foods and Packaging Materials

I. INTRODUCTION

One nonthermal method of food preservation involves the use of intense and short-duration pulses of broad-spectrum "white" light. The technology of using light pulses is applicable mainly in sterilizing or reducing the microbial population on the surfaces of packaging materials, on packaging and processing equipment, foods, and medical devices, as well as on many other surfaces (1). Traditionally, the packaging material used in aseptic processing is sterilized with hydrogen peroxide. Residues of hydrogen peroxide in the packaging material or in the food product may be highly undesirable (2). Light pulses may be used to reduce or eliminate the need for chemical disinfectants and preservatives.

The spectrum of light used for sterilization purposes includes wavelengths in the ultraviolet (UV) to wavelengths in the near infrared region. The material to be sterilized is exposed to at least one pulse of light having an energy density in the range of about 0.01 to 50 J/cm^2 at the surface, using a wavelength distribution such that at least 70% of the electromagnetic energy is distributed in a wavelength range from 170 nm to 2600 nm (3). The *PureBright*™ process developed by PurePulse Technologies, Inc., San Diego, Calif., utilizes light with an intensity of about 20,000 times that of sunlight at the surface of the earth. The *PureBright* spectrum includes wavelengths between 200 and 300 nm that are not present in sunlight filtered by the atmosphere surrounding earth. Inactivation of select resistant microorganisms requires treatment with the complete spectrum; other microorganisms are inactivated with a filtered spectrum. The filtered spectrum includes light in a certain wavelength range only.

The pulse light process uses short-duration flashes of broad-spectrum white light to inactivate a wide range of microorganisms, including bacterial and fungal spores (1). The duration of pulses ranges from 1 μs to 0.1 s (4). The flashes are typically applied at a rate of 1 to 20 flashes per second. For most applications, a few flashes applied in a fraction of a second provide high level of microbial inactivation (1). Therefore, the process is very rapid and amenable to high throughput.

II. GENERATION OF LIGHT PULSES

PureBright uses a technique known as pulsed energy processing (1). By storing electrical energy in a high energy density electrical storage capacitor and releasing it in short, high intensity pulses, high peak power levels can be generated. Such high peak power pulses of electrical energy can be used to create intense pulses of light or pulses of high electric field. Stored energy pulses an inert gas lamp to produce an intense flash of light lasting only a few hundred microseconds (5). The short, intense light pulses or flashes may be generated by pulsed gas-filled flashlamps, spark gap discharge apparatus, or other pulsed light sources. Pulsed gas-filled flashlamps produce broad band light when a pulse of electrical current is discharged through the flashlamp. The current ionizes the gas, resulting in a broad-spectrum white light flash containing wavelengths from 200 nm in the UV to 1 mm in the near infrared (5). Approximately 25% of the emitted light is UV, 45% visible, and 30% infrared (1). Such flashlamps typically employ inert gases such as xenon or krypton because of the high efficiency of electrical to optical energy conversion (2).

The *PureBright* system (Fig. 1) for generating light pulses consists of two main parts: the power unit and the lamp unit. The power unit generates high voltage, high current pulses used to energize the lamps. The unit operates by converting line voltage AC power into high voltage DC power. The high voltage DC is used to charge a capacitor. Once the capacitor is charged to a preset point, a high voltage switch discharges the light energy from the capacitor to the lamps (Fig. 2). The lamp-firing sequence may be controlled by an internal controller or by interfacing with the packaging/processing machine controller (4).

The lamp unit consists of one or more inert gas lamps arranged to illuminate a desired treatment area. The lamp unit is connected to the power unit by a high voltage cable. To flash the lamp, a high voltage, high current pulse is ap-

FIGURE 1   PBS Series *PureBright* System. (Courtesy of PurePulse Technologies, Inc.)

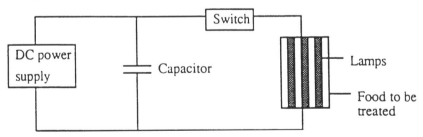

FIGURE 2   Simplified version of a system for treating foods with intense light pulses.

plied. The high current passing through the gas in the lamp emits an intense pulse of light. The flashing frequency is selected for the particular processing or packaging line. More than one lamp may be used, and the lamps may be flashed simultaneously or sequentially. Passage of the product through a trigger sensor zone may be utilized to control the flashlamp pulses (4).

III. THEORY

Treatment of products by using short duration intense pulses of light is based on the penetration of light through the product. Light penetrates the surface of materials according to the equation given by (2):

$$I = (1 - R) I_o e^{-x} \tag{1}$$

where I is the energy intensity of light transmitted to a distance below the surface, R is the surface coefficient of reflection, I_o is the intensity incident upon the surface, and x is the extinction coefficient, a determination of the opacity of the material.

The light that penetrates the material but is not transmitted is dissipated as heat. The heat energy dissipated per unit area, E_d, at depth d, is described by (2):

$$E_d = (1 - R) I_o (1 - e^{-d}) \tag{2}$$

As the result of the temperature gradient between the surface and the inner layers of the material, energy dissipated as heat is transferred to the inner layers of the material by thermal conduction. Heat transferred by conduction, E_c, is described by (2):

$$E_c = A k t \frac{dT}{dx} \tag{3}$$

where A is the area of exposure, k is thermal conductivity of the material, t is time, dT is the temperature difference, and dx is the heat transfer distance.

Heat transfer continues until a steady state is attained and the temperature of the bulk of the material becomes constant. The time required to reach a constant temperature depends on the amount of heat dissipated and the thermal properties of the material. Because food often contains water, a good thermal conductor, surface heat is rapidly conducted into the food. However, if the duration of the light pulses is short compared to the thermal conductivity time constant, the energy from the pulses may be deposited at the surface for a very short time, during which little or no thermal conduction occurs. This results in substantially instantaneous heating of a thin surface layer to a temperature much higher than the steady-state temperature achieved by a continuous light beam of equivalent mean power (1). Various standard fluids such as air and water exhibit a high degree of transparency to a broad range of wavelengths including visible and UV light. Other liquids, such as transparent sugar solutions and wines, may exhibit a more limited transparency. For satisfactory disinfection, it is preferable that the fluid have a transparency to UV light defined as follows: at least half of the incident light at 260 nm is transmitted through 0.25 cm into the fluid. When the material is opaque, very little light penetrates the material, and substantially all the light is dissipated within a very superficial surface layer—typically less than 1 mm thick—of the foodstuff (3).

IV. EQUIPMENT

Figure 3 represents an aseptic packaging apparatus containing a reel of conventional flexible aseptic packaging material guided by a series of rollers through a solution of an absorption-enhancing agent. The absorption-enhancing agent is optional and used only when required. The packaging laminate material composed of one or more internal coating and sealing layers is dipped in the enhancing agent (2).

When the surface of the material to be treated is opaque, or the light absorption coefficient is low, absorption-enhancing agents have to be utilized. The enhancing agents may be applied by spraying, dusting the surface with a powder containing the agent, applying the agent as a dissolved liquid, or vaporizing the agent onto the surface of the packaging material. A suitable absorption-enhancing agent will exhibit a high optical absorption coefficient at the desired spectral wavelength. Although the absorption-enhancing agent may be completely removed from the product during processing, food products and medical devices require an edible enhancing agent. The enhancing agent is selectively absorbed on the cell surface. Photon-sensitive indicator agents such as dyes, pH sensitive or sensitive to oxidation potential, are utilized as absorption-enhancing agents. The absorption coefficient of some dyes changes during the light pulse treatment. FDA-approved colors such as carotene, red dye number 3, lime green, black cherry, and mixtures thereof may be used as absorption-enhancing agents. Natural or cooking oils may also be used as enhancing agents. Mixtures of two or more components having different absorption maxima may be used to increase the optical absorption over the desired spectrum (2).

The use of absorption-enhancing agents also allows the pulse duration

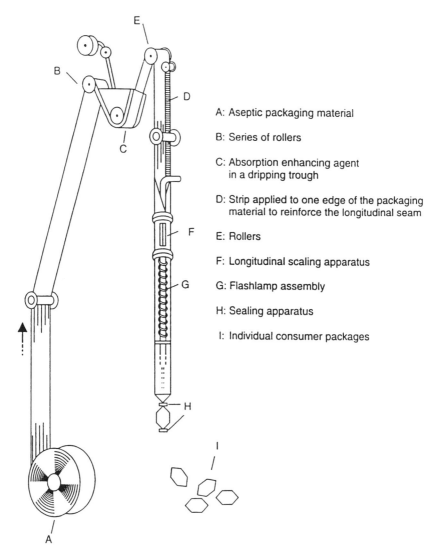

A: Aseptic packaging material

B: Series of rollers

C: Absorption enhancing agent
in a dripping trough

D: Strip applied to one edge of the packaging
material to reinforce the longitudinal seam

E: Rollers

F: Longitudinal scaling apparatus

G: Flashlamp assembly

H: Sealing apparatus

I: Individual consumer packages

FIGURE 3   Aseptic packaging apparatus using light to sterilize the packaging material. (Ref. 2, US Patent 5,034,235)

of the light to be increased. Light pulses with long duration exhibit the same sterilization effect as short pulses, but at a reduced UV content in the light. A reduced UV content requirement in light will increase the lifetime of the flashlamps (2).

Excess absorption agent is removed by the rollers and the film is subsequently formed into a longitudinally sealed tube by the sealing apparatus (2).

The product filling and flashlamp apparatus are shown in detail in Figure 4. The flashlamp apparatus is composed of an outer support tube with one or more flashlamps attached to the tubular support. The flashlamps are distributed along the tubular support so that upon pulsing, the entire inner surface of the sealed packaging material tube is subject to intense short-duration pulses. The flashlamps can be arranged in several ways along the tubular support so

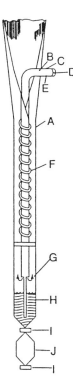

A: Flashlamp assembly

B: Outer support tube

C: Flashlamp electrical cable

D: Sterile food product tube

E: Lamp coolant lines (optional)

F: Flashlamp

G: Sterile air moving upward

H: Sterile food product

I: Sealing apparatus

J: Individual consumer package

FIGURE 4   A product filling and flashlamp apparatus. (Ref. 2, US Patent 5,034,235)

that the entire surface of the packaging material tube is exposed to the light pulses. In operation, the longitudinally sealed tube is filled with commercially sterile food product. The tube is then advanced one package length and subjected to the desired number of pulses to sterilize the area adjacent to the food product. Sterile air is used to cool the flashlamps, remove from the tube any photochemical products produced by pulses, and prevent contamination from settling on the treated area (2).

Aseptic packaging apparatus using preformed containers are also available (Fig. 5). The containers may be optionally sprayed with the absorption-enhancing agent and subjected to the light pulses (Fig. 6). The containers are passed through several stations while being disinfected. The disinfected containers are filled with a processed, commercially sterile food product. The

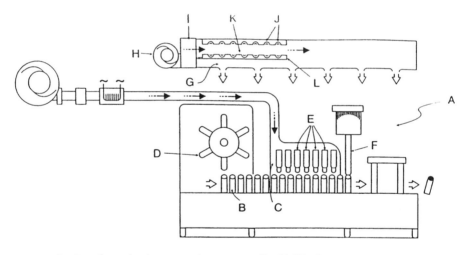

A: Aseptic packaging apparatus
B: Preformed product containers
C: Sterilization zone
D: Absorption enhancing agent
   spraying apparatus (optional)
E: Pulsed light treatment station
F: Product filling station

G: Air filtration apparatus
H: Air input blower
I: Filter
J: Flashlamps
K: Pulsed light treatment zone
L: Reflective housing

FIGURE 5 Aseptic packaging apparatus using light-sterilized preformed product containers. (Ref. 2, US Patent 5,034,235)

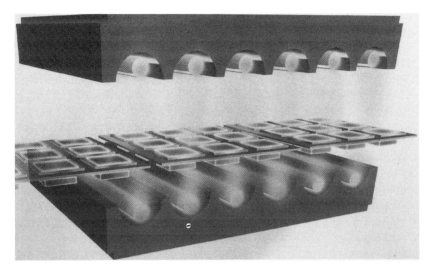

FIGURE 6 *PureBright* light can be transmitted through many types of clear packaging. (Courtesy of PurePulse Technologies, Inc.)

containers are subsequently sealed at the top with a sterile lid. The lids may also be disinfected using light pulses. A laminar sterile air curtain may be provided over the entire aseptic packaging apparatus to prevent the contamination of the packaged units (1).

Figure 7 describes an apparatus used for sterilizing pumpable foods such as water or fruit juices. The apparatus consists of a reflective cylindrical enclosure defining a treatment chamber. The food product flows through the chamber surrounding a pulsed light source. A liquid circulation pump controls the flow rate of the product through the chamber according to the pulse repetition rate so that the food product is subjected to the selected number of pulses. The treatment chamber may be designed to include a reflector assembly as an outer wall or external reflector to reflect back the illumination traversing the food product. Fluids such as air and water are relatively transparent to light, causing little or no attenuation in the flux density. However, for fluids with significant absorption, significant reduction in flux density occurs. Therefore, maintenance of a minimum flux density throughout the treatment zone and mixing must occur to ensure that the entire fluid is subjected to the appropriate flux density (2).

A: Treatment chamber

B: Pulsed light source

C: Liquid circulation pump

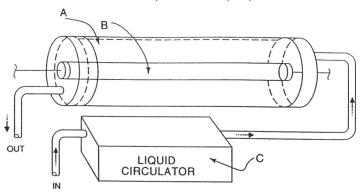

FIGURE 7   Apparatus used for sterilizing pumpable products. (Ref. 2, US Patent 5,034,235)

V. APPLICATIONS OF LIGHT PULSES

Light pulses can inactivate microorganisms on packaging material used in aseptic processing, liquid foods, solid foods such as fish and meat products, and baked foods. The *PureBright* system can reduce the vegetative microorganism population by about nine log cycles, and the spore population can be reduced by seven log cycles on a smooth, nonporous surface. On porous and complex surfaces such as meat, an approximately two to three log cycle reduction is obtained. It has also been observed that mold spores rather than bacterial spores are more resistant to UV light (4). In the case of meat products, thin slices will permit light penetration through the food material. In addition, prepared and processed meat products such as sausages and ground meat patties could be treated to increase their shelf life under refrigeration without the necessity for freezing. Similarly, vegetables such as potatoes and tomatoes, fruits such as apples and bananas, and prepared food products such as pastas and rice entrees can be treated to increase shelf life. Tomatoes are very susceptible to surface mold formation after a few days, even when stored under refrigeration. Fresh tomatoes treated at PurePulse Technologies with pulsed light and stored under refrigeration remained acceptable for 30 days.

Similarly, white bread slices treated through the packaging material maintained a fresh appearance for more than 2 weeks, whereas untreated white bread slices became moldy. Exposure to light pulses reduces the populations of *Listeria* and *Salmonella* inoculated in meat, with minimal changes in the nutrient content (6).

Food surfaces and packaging materials are typically exposed to between 1 and 20 pulses of high intensity, short duration light. The short duration of each pulse permits spatial localization of the lethal effects of light pulses to a thin layer, such as the surface of the product.

The number of lamps, flashing configuration, and pulse rate depend on the type of product and extent of treatment required. If the intensity of light or number or duration of pulses is high, the increase in temperature of the product may be greater than desirable. The intensity of light should be sufficient to heat a superficial layer of the product having a thickness of less than 10 μm to at least 50° to 100°C. Heat may be localized at the superficial surface layer without significantly raising the interior temperature. The number of pulses and total energy may be limited to maintain the surface temperature below 50° to 100°C ten seconds after the pulsed light exposure. Even though heat is generated in the product when exposed to light, it is very small compared to the amount of heat necessary for it to be adequately processed thermally (3).

Food products can be disinfected by pulsed light after they are placed in a packaging material that is sufficiently transparent to the treatment spectrum. The packaging material must transmit at least 10–50% of the light energy over a predetermined treatment wavelength range of less than 320 nm (3). Foods can be disinfected using the full spectrum or selected spectral distributions for particular microorganisms. Filtering of the spectrum eliminates undesirable wavelengths that adversely affect food flavor and quality. Filtering is accomplished with solid filters such as UV-absorbing glass filters or with liquid filters formed by a liquid either standing in or flowing through a jacket surrounding a flash lamp. Organic or inorganic liquids absorb wavelengths to be eliminated.

Fluence is a measure of the incident light energy per unit surface area (J/cm²). It is desirable for certain aseptic packaging processes that the packaging material be treated with light pulses having a relatively high UV content to minimize the total fluence necessary to achieve the desired reduction in microbial population (3).

The biological effects of the UV wavelengths contained in the pulse light spectrum are well documented (5) The antimicrobial effects of UV wavelengths are primarily mediated through absorption by highly conjugated carbon-to-carbon double-bond systems in proteins and nucleic acids. Comparison of the antimicrobial effects obtained using pulsed light with those obtained using non-pulsed or continuous wave conventional UV sources shows a significantly higher inactivation for pulsed light. More than 7 logs of *Aspergillus niger* spore inactivation result with few pulsed light flashes. In the case of continuous wave UV light, after 3 to 5 logs of inactivation "tailing" is observed; the process becomes very inefficient; and large increases in exposure time (energy) produce little to no improvement in inactivation (5).

A variety of microorganisms, including *Escherichia coli, Staphylococcus aureus, Bacillus subtilis*, and *Saccharomyces cerevisiae*, were inactivated by using between 1 and 35 pulses of light with intensity ranging between about 1 and 12 J/ cm² (Figs. 8 through 17). Greater inactivation is obtained when full-spectrum light rather than glass-filtered-spectrum light is used. Thus, the UV component of light is needed to inactivate microorganisms (3).

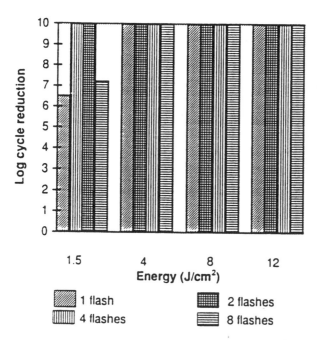

FIGURE 8   Inactivation of *Escherichia coli* using full-spectrum light. (Ref. 2, US Patent 5,034,235)

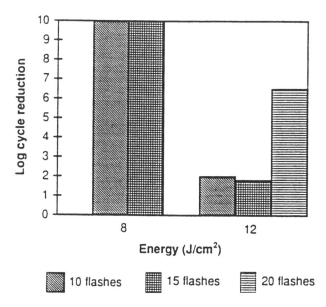

FIGURE 9 Inactivation of *Escherichia coli* using filtered-spectrum light. (Ref. 2, US Patent 5,034,235)

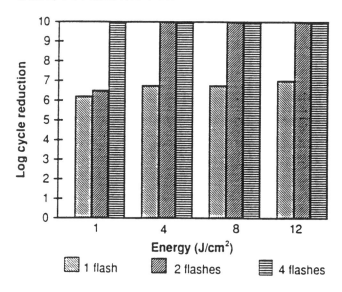

FIGURE 10 Inactivation of *Bacillus subtilis* vegetative cell susing full-spectrum light. (Ref. 2, US Patent 5,034,235)

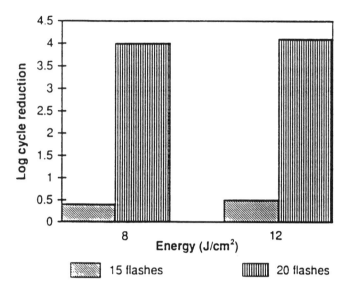

FIGURE 11 Inactivation of *Bacillus subtilis* vegetative cells using filtered-spectrum light. (Ref. 2, US Patent 5,034,235)

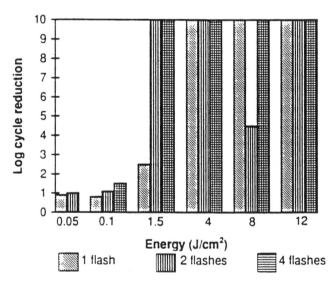

FIGURE 12 Inactivation of *Bacillus subtilis* spores using full-spectrum light. (Ref. 2, US Patent 5,034,235)

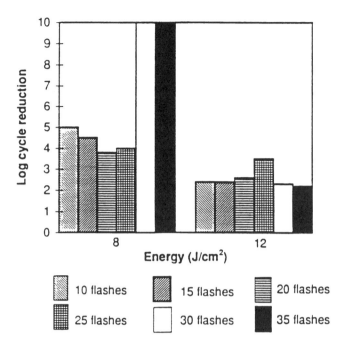

FIGURE 13 Inactivation of *Bacillus subtilis* spores using filtered-spectrum light. (Ref. 2, US Patent 5,034,235)

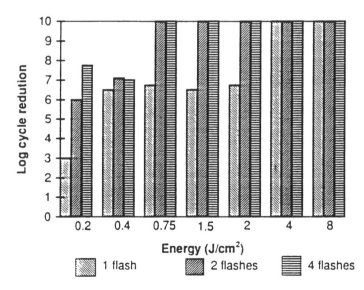

FIGURE 14 Inactivation of *Staphylococcus aureus* using full-spectrum light. (Ref. 2, US Patent 5,034,235)

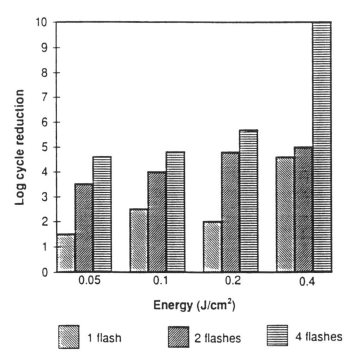

FIGURE 15   Inactivation of *Saccharomyces cerevisiae* using full-spectrum light. (Ref. 2, US Patent 5,034,235)

Shrimp treated with pulsed light and stored under refrigeration for 7 days remain edible whereas untreated shrimp show extensive microbial degradation and are discolored and foul smelling (5). Treating raw shrimp with four to eight flashes of a polychromatic light with a fluence of 1–2 J/cm² resulted in a shelf-life extension of one week, compared to untreated control shrimp. Exposure of shrimp inoculated with *Listeria* and chicken inoculated with *Salmonella* to light pulses resulted in a one to three log cycle reduction in the bacterial population (3).

Extensive tests on a variety of meats have shown that pulsed light can be used to enhance product shelf life and safety (5). With the *PureBright* process, *Salmonella* can be reduced by two log cycles on chicken wings, in samples inoculated with either 5 or 2 logs/cm². *Listeria* was reduced by two log cycles on hot dogs (inoculated with 3 or 5 logs/wiener) after pulsed light treatment. A two log cycle reduction is also obtained for pulsed light–treated

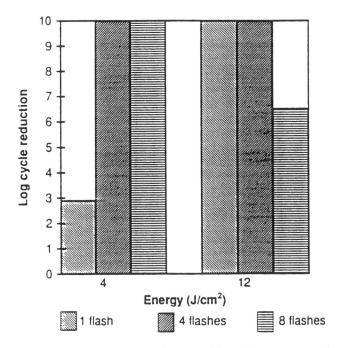

FIGURE 16 Inactivation of *Aspergillus niger* spores using full-spectrum light. (Ref. 2, US Patent 5,034,235)

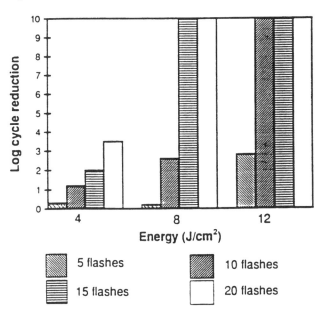

FIGURE 17 Inactivation of *Aspergillus niger* spores using filtered-spectrum light. (Ref. 2, US Patent 5,034,235)

primal and retail beef cuts. The vacuum-packaged beef exhibits no change in color, appearance, or taste after $2^1/_2$ weeks. The AMI Foundation of the American Meat Institute is currently evaluating the *PureBright* process as part of a hurdle system in combination with methods such as acetic acid spray and hot water washes to reduce pathogens on the eviscerated carcasses (7).

Curds of commercial dry cottage cheese were inoculated with *Pseudomonas* and subjected to light with an energy density of 16 J/cm² and pulse duration of 0.5 ms. After two flashes, the viability of microbial population was reduced by 1.5 log cycles. The temperature at the surface of the curd closest to the light source increased by 5°C. Sensory evaluation using experienced taste testers showed no effect on the taste of cheese as a result of the pulsed light treatment (3).

Fresh fish is susceptible to microbial deterioration resulting in undesirable odors and flavors. The technology of disinfection of food products using light pulses may be combined with other methods of sanitation. For example, fresh fish may be washed with high pressure water and exposed to light pulses. The fish may be washed with sterile water after exposure to light to remove the surface heated layer. A combination of high pressure wash and exposure to pulsed light reduced the psychrotroph and coliform population on the surface of fish tissue by 3 log cycles. The fish samples were sensory acceptable for 15 days of refrigerated storage (2).

Commercial eggs off the shelf of a grocery store and raw unwashed eggs from a farm were inoculated with *S. enteritidis* by immersion for 10 min and treated with eight flashes at 0.5 J/cm² per flash (1). Pulsed light was very effective in eliminating microbial contamination from the surface of shell eggs. As many as eight log cycle reductions were obtained. No difference between commercial and raw eggs was noted. The inactivation effect on shell eggs is not limited to surface microorganisms but also extends to a degree into the egg pore, as a second set of experiments demonstrated (1).

Vegetables and fruits such as potatoes, bananas, and apples can undergo enzymatic browning. Browning is caused by polyphenol oxidase (PPO) and flavonoid substrates present in many fruits and vegetables. PPO can be inactivated using light pulses. Slices of potatoes cut and exposed on one side to pulsed light retain a fresh appearance. The unexposed surface, however, does undergo browning. The activity of alkaline phosphatase, which catalyzes the hydrolysis of phosphatase esters, is reduced by 60–70% with a single full spectrum flash of light at fluence of 1 J/cm². However, a filtered-spectrum flash of light is less effective in reducing the activity of alkaline phosphatase (Figs. 18 and 19). The inactivation of enzymes is related to photochemical effects (3).

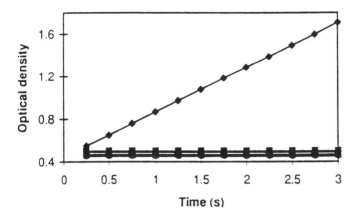

FIGURE 18 Inactivation of alkaline phosphatase using two flashes of full-spectrum light. (Ref. 2, US Patent 5,034,235)

| ♦ control | ■ 1 J/cm^2 | ▲ 2 J/cm^2 |
| × 3 J/cm^2 | * 4 J/cm^2 | ● 6 J/cm^2 |

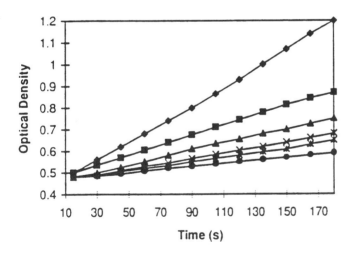

FIGURE 19 Inactivation of alkaline phosphatase using two flashes of filtered-spectrum light. (Ref. 2, US Patent 5,034,235)

| ♦ control | ■ 1 J/cm^2 | ▲ 2 J/cm^2 |
| × 3 J/cm^2 | * 4 J/cm^2 | ● 6 J/cm^2 |

The surface of different packaging materials was inoculated with *S. au-reus, B. cereus* spores, and *Aspergillus niger* spores in a concentration of 10 to 1000 CFU/cm². The surfaces of the inoculated packaging material were exposed to a single light pulse with a duration of 1.5 ms at select intensities. *S. aureus* was inactivated with an intensity as small as 1.25 J/cm²; *B. cereus* and *Aspergillus* spores were inactivated with intensities greater than 2 J/cm². Mold spores are more resistant to pulsed light than are bacterial spores (3).

A method for increasing shelf life of perishable solid food products by applying a solid edible coating that is transparent to light, after which the product is illuminated with light to deactivate surface microorganisms, has been patented (8). Strawberries did not exhibit surface mold growth after 2 weeks at ambient air and temperature when wax-coated and exposed to four pulses or flashes having 0.5 J/cm² energy density and 0.3 ms of duration with a repetition rate of 1 pulse/s.

High levels of inactivation of bacterial spores, *Cryptosporidium* oocysts, *Klebsiella terragena*, and viruses have been demonstrated in water (1). Applications being investigated include treatment of municipal water, waste water, and packaged drinking water before, during, or after packaging (5).

Very high levels of microorganisms are inactivated on surfaces and in transparent media such as water that allows the light to penetrate. On the generally opaque and irregular surfaces of foods, some microorganisms may be shielded from the pulsed light. However, 1–3 log cycles of total microbial reduction are achieved even on the highly irregular surface of meats, resulting in significant shelf life extension and pathogen reduction (5). A petition was submitted to the Food and Drug Administration in February 1994 for approval of pulsed light food treatment. Pulsed light treatment costs are very favorable (5). Recently, PurePulse Technologies partnered with Tetra Laval to commercialize the *PureBright* process. The treatment costs for pulsed light at 4 J/cm², including equipment amortization, lamp replacement, electricity, and maintenance are estimated at 0.1 cent per square foot of treated area (7).

In addition to disinfection of foods, intense light pulses may also be used for treatment of

- Cosmetics and ingredients used in manufacturing cosmetics
- Equipment, products, devices or areas requiring a high degree of cleanliness, e.g., aseptic processing
- Medical and dental instruments prior to use
- Food processing equipment, to reduce the levels of contamination and possibility of cross-contamination

- Processed or partially processed sewage effluents, to reduce the microorganism or viral burden
- Air or other gases or gaseous compounds, to reduce the microorganism burden.

VI. FINAL REMARKS

The technology of utilizing short pulses of light is an attractive alternative for disinfecting packaging materials and food products packaged in transparent films. Heat generated in the light pulse process is minimal, and microorganisms are inactivated by a combination of photochemical and photothermal mechanisms. Light pulses apparently do not affect the nutrient retention in foods, although a detailed study is not available. Studies of the effects of light pulses on the properties of food beyond food safety and spoilage are necessary. Assessing the energy efficiency of light pulses compared to thermal and other nonthermal processes will also be beneficial.

REFERENCES

1. J. Dunn, Pulsed light and pulsed electric field for foods and eggs. *Poultry Sci.* 75: 1133-1136 (1996).
2. J. E. Dunn, R. W. Clark, J. F. Asmus, J. S. Pearlman, K. Boyer, and F. Pairchaud, Methods and apparatus for preservation of foodstuffs. Int. Pat. Appl. No. WO 88/03369 (1988).
3. J. E. Dunn, R. W. Clark, J. F. Asmus, J. S. Pearlman, K. Boyer, F. Pairchaud, and G. Hofman, Methods and apparatus for preservation of foodstuffs. U. S. Pat. 5,034,235 (1991).
4. Anonymous, The *PureBright* Process. PurePulse Proprietary Information (1994).
5. J. Dunn, T. Ott, and W. Clark, Pulsed light treatment of food and packaging. *Food Technol.* 49(9): 95-98 (1995).
6. J. Rice, Sterilizing with light and electrical impulses *Food Proc.* July, pp. 66 (1994).
7. Anonymous, PurePulse teams up with Tetra Laval to commercialize *PureBright*. *Food Eng.* January, pp. 24-25 (1995).
8. J. E. Dunn, T. M. Ott, and R. W. Clark, Prolongation of shelf-life in perishable food products. U. S. Pat. 5,489,442 (1996).

7

Food Irradiation

I. INTRODUCTION

Food irradiation is a topic attracting much interest and comment—unfortunately, much of it inadequately informed. Therefore, one of the main chal-·
lenges for the wide range of interested individuals in the food world is to be
well informed on this preservation process.

Food irradiation is a "cold" process for preserving food that has been
studied extensively for over 45 years. Beneficial results, including extension
of the storage life of root crops; disinfestation of spices, fruit, and cereals; reduction of the population of many microorganisms responsible for spoilage;
delay of fruit ripening; improvement of sensory properties in some foods; and
the destruction or reduction of the largely unavoidable pathogens that contaminate foods, especially raw food of animal origin, can be achieved without
the production of radioactivity or toxic compounds, and with no vital loss in
the nutritional quality of the foodstuff. The beneficial aspects of this process
are coupled with safety issues that are the subject of many political debates
and public scrutiny, stimulated to a large extent by individuals who are often
misinformed, and sow confusion, as well as by certain mature and respected
consumer advocacy groups.

On the basis of an international consensus, in 1983 the Joint Food and
Agriculture Organization/World Health Organization Codex Alimentarius
Commission accepted food irradiation as a safe and effective technology for
the treatment of food and adopted a Codex General Standard for Irradiated

Foods with an associated Code of Practice. Although the use of irradiation continues to grow worldwide, negative reaction in various countries restricts its common use. Introduction into commercial practice is somewhat slow because many governments require comprehensive data to support the wholesomeness of irradiated food, and there are lengthy regulatory procedures for granting approval for the food or process, often with the involvement of the public. Here, as in many other instances, the rejection of the process is based primarily on lack of information. Besides many educational efforts in recent years, the widespread information campaigns demanded by the World Health Organization are largely needed.

This chapter is an overview of food irradiation, summarizing the progress made, pointing out some of the advantages and limitations of irradiation, and indicating its potential for the future. There are extensive compilations of 45 years' wealth of information published on the subject of irradiation. Some examples include the three volumes of *Preservation of Food by Ionizing Radiation*, edited by Josephson and Peterson (1); the comprehensive reports of the Council for Agricultural Science and Technology, edited by Wierbicki et al. (2,3); the notable irradiated dry ingredients anthology by Farkas (4); and the great synopsis in *Food Irradiation* of Urbain (5).

II. HISTORICAL ASPECTS

Food irradiation is not, as often described by media reports, a new technology. In fact, irradiation of food is much older and was studied more thoroughly than other processing and preservation innovations introduced by the food industry in recent years. Irradiation is widely used around the world to sterilize medical supplies (6). The discovery of x-rays by W. K. Roentgen in 1895 and radioactive substances by H. Becquerel a year later led to intense research on the biological effects of ionizing radiation. Irradiation was first patented for food preservation in 1905 by two British scientists. Food irradiation was first used in the United States in 1921 to inactivate the human parasite *Trichinella spiralis*, which contaminated pork muscle (7). The biomedical literature of the 1920s and 1930s contains numerous articles on the effects of x-rays on foods and food constituents.

International research intensified in the 1940s, when irradiation was considered a means of providing safe food for troops in remote areas, and in the 1950s, when relatively large quantities of inexpensive artificial radioisotopes became available (6,8,9). In the United States, food irradiation was approved in 1963 for control of insects in wheat and wheat flour (10). United States astronauts have been eating irradiated food since 1972, when the

Apollo 17 crew selected ham for the first irradiated flight meal. Immunocompromised people also benefit from irradiated food because it helps reduce the risk of bacterial infection (8). In 1987, the European Economic Community, except for Great Britain and West Germany, approved the irradiation process for specified foods (11).

Today, more than 40 countries (Table 1) have approved more than 100 irradiated food items or groups of food for consumption, either on an unconditional or restricted basis (12–14); of these countries, 25 have approved commercial production of some foods. Most of the countries irradiate spices to inactivate bacteria and molds. Other commonly irradiated products include potatoes and onions (for prevention of sprouting) (15), cereals and flour, fresh fruit, and poultry (9). There is a worldwide research base of 9000 references for food irradiation that covers aspects of safety and wholesomeness for a wide range of produce, grain, fish, meat, and poultry (8). Traditional processing techniques—at least those that do not adversely affect sensory qualities— are inadequate to render foods free of pathogenic bacteria. However, gamma irradiation can be used to achieve this goal (16). Irradiation may prove to be one of the most important food technologies for the protection of human health since the introduction of pasteurization of milk nearly a century ago (17) and, as stated by Urbain (18), can be a giant step beyond Appert.

TABLE 1 Countries Approving Food Irradiation

Algeria	Germany	Philippines
Argentina	Hungary	Poland
Bangladesh	India	Russia
Belgium	Indonesia	South Africa
Brazil	Iran	Spain
Bulgaria	Israel	Syria
Canada	Italy	Thailand
Chile	Ivory Coast	Ukraine
China	Japan	United Kingdom
Croatia	Korea	United States
Cuba	Mexico	Uruguay
Czech Republic	Netherlands	Vietnam
Denmark	New Zealand	Yugoslavia
Finland	Norway	
France	Pakistan	

Source: From Refs 13 and 14.

III. SOME TECHNOLOGICAL ASPECTS OF FOOD IRRADIATION

Food irradiation is a physical means of food processing that involves exposing prepackaged or bulk foodstuffs to gamma rays, x-rays, or electrons.

A. Radiation Sources

Foodstuffs are generally irradiated with gamma radiation from a radioisotope source, or with electrons or x-rays generated using an electron accelerator. Similar irradiation sources are used for the irradiation sterilization of disposable medical products. Over 170 commercial irradiation facilities are in operation around the world to irradiate medical supplies. About one-third of the facilities operate as multipurpose facilities that also treat some foodstuffs. Although different equipment can be used to produce the different ionizing radiations, the same chemical changes are produced by the different ionizing radiations. The only practical differences relate to their powers of penetration and hence the dimensions and density of the food product capable of being irradiated (19). The radioisotope used in most commercial gamma irradiation facilities is cobalt 60, the greater part of which is produced by irradiating natural cobalt (^{59}Co) in Canada (Fig. 1 shows a commercial ^{60}Co irradiator). Cesium 137 from spent nuclear fuel is an alternative radioisotope source, though it is not used to any great extent at present. The cost of gamma irradiation is competitive with other methods of food processing.

Commercial electron accelerators produce beams of electrons at very high power levels; at the electron energies accepted for food irradiation, however, the electrons do not penetrate deeply in the irradiated material. The electron beam can be stopped in a heavy metal target to produce x-rays, which are more penetrating, but in the process a large part of the beam energy is lost as heat in the target. For commercial facilities with high throughput, total investment and per unit cost breakdown are approximately the same for the electron beam and cobalt 60. Electron accelerators are generally employed in irradiation processing to treat relatively thin layers of material. For reason of quality control, electron irradiation is preferably applied from one side only. The reason for the preference is that when the product is irradiated from one side only, it is possible for a detector behind the product to monitor and continuously assure that the beam is penetrating the product. The maximum practical product thickness is only about 3.3 cm of unit density material such as water (20). Electrons and x-rays are restricted to maximum energies of 10 and 5 MeV, respectively, in food irradiation applications, to limit the extent of nu-

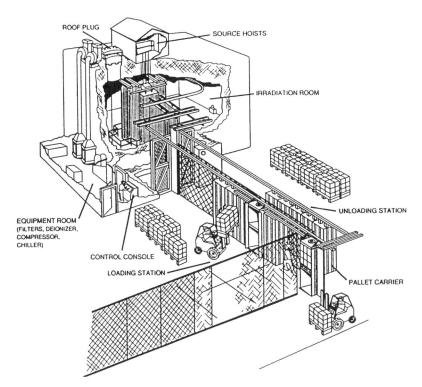

FIGURE 1 Commercial cobalt 60 automatic pallet irradiation facility. (Courtesy of Nordion International, Inc.)

clear reactions that induce radioactivity in the irradiated product. The mean energy of cobalt 60 gamma radiation, 1.25 MeV, is below this limit. Research concludes that at these energies, radioactivity induced is no greater than the natural radioactivity of the foodstuff caused by ^{14}C and ^{40}K; and, furthermore, the induced activity decays rapidly, by a factor between 10 and 20 during the first 24 h following irradiation. The natural radioactivity does not decay significantly over the normal storage life of the foodstuff (9). Electron beam facilities have an economic advantage over cobalt 60 irradiators if the thickness of the product is small, throughput is large, and integration into the production line is possible (17,20). Further, the x-ray source can be shut off while not in use, but the cobalt 60 will continue to irradiate and decay even when not in use.

The size of the source increases with the irradiating dose and the throughput rate, and is given by (20):

$$S = \frac{XD}{3600F} \tag{1}$$

where S is the source size (kW), X is the throughput rate (kg/h), D is the dose (kGy), and F is the fraction of the radiation energy that is usefully absorbed. F depends on the design of the irradiation geometry, the product form and density. For many applications the values of F may be 0.40, 0.25, and 0.05 for an electron accelerator, cobalt 60, or x-ray facilities, respectively [20]. The size of cobalt 60 sources is usually measured in curies or becquerels (Bq): 1 kW = 67 300 curie = 2.49 x 10^{15} Bq are the conversion factors.

Comparative studies on the effects of gamma rays and electron beams are quite important for designing irradiation plants and establishing irradiation conditions for treating foods. Based on the results of numerous studies on the comparative effects of gamma rays and electron beams, or of high dose rates of irradiation or low dose rates of irradiation, the biological effects of electron beams on microorganisms and insects are slightly lower than those of gamma rays, although the difference in the effect between the two types of irradiation is influenced by various parameters, including oxygen concentration and water content. The difference in the effect on chemical reactions in foods between the two types of irradiation is much smaller than the difference observed on the biological effects (21).

B. Absorbed Dose

The degree of chemical and physical change produced when food is exposed to high energy irradiation is determined by the energy absorbed. In irradiation processing, it is described as the absorbed dose, or simply dose, and is measured in units of kilogray (kGy) where one gray (Gy, after L. H. Gray) is an energy absorption of one joule per kilogram. The rad (1 rad = 0.01 Gy) was widely used in the past. Special terms that are sometimes used to describe the applications of irradiation in food processing (5,9,22) follow.

Radicidation

Radicidation is the treatment of food with a dose of ionizing radiation sufficient to reduce the number of specific viable non-spore-forming pathogenic bacteria to such a population that none is detectable in the treated food when examined by recognized bacteriological testing methods; radicidation may also be applied to inactivation of parasites. It is a treatment with relatively low doses (about 0.1–8 kGy) to eliminate pathogenic organisms and microorganisms other than viruses. Radicidation destroys organisms such as tapeworms and *Trichina* in meat (requiring about 0.1–1 kGy) and decreases the number of viable non-spore-forming pathogenic microorganisms (requiring about 2–8 kGy). Radicidation refers to "irradiation pasteurization" treatments, particularly with the intent of eliminating a specific pathogen.

Radurization

Radurization is the treatment of food with a dose of ionizing radiation sufficient to enhance its keeping quality by causing a substantial reduction in the numbers of specific viable spoilage microorganisms. It is a treatment with doses of about 0.4–10 kGy to improve the shelf life of the product. Radurization also refers to "irradiation pasteurization."

Radappertization

Radappertization is the treatment of food with a dose of ionizing radiation sufficient to reduce the number and/or activity of viable microorganisms (viruses excepted) to such a level that very few, if any, are detectable by recognized bacteriological or mycological testing methods applied to the treated food. Radappertization treatment must be such that no spoilage or toxicity of microbial origin is detectable no matter how long or under what conditions the food is stored after treatment, provided the food is not recontaminated. It is an irradiation treatment with doses of about 10–50 kGy to bring about virtually complete sterilization. Radappertization is a term used to define "irradiation sterilization" or "commercial sterility" as is understood in the canning industry, with the resulting product being shelf-stable under normal conditions.

The Joint Expert Committee on the Wholesomeness of Irradiated Foods convened by the World Health Organization (WHO), Food and Agriculture Organization (FAO), and the International Atomic Energy Agency (IAEA) pronounced food irradiation safe in 1970. The Committee concluded in 1980 that the irradiation of food commodities up to an overall mean dose of 10 kGy causes no toxicological hazard and introduces no special nutritional or microbiological problems. The 10 kGy dose level is not an upper limit of safety but simply a level to which safety has been proved (9). Numerous expert committees are evaluating and reevaluating the results of these investigations (13) and confirm the safety of ionizing irradiation of food. In sum, as pointed out by Pauli (23), while it is important to evaluate technological changes in food processing carefully, there is no doubt that irradiation of food has successfully surpassed such an evaluation. Studies evaluating high dose ionizing irradiation of food in the United States conclude a dose of 58 kGy will not produce deleterious effects (6).

IV. ADVANTAGES AND LIMITATIONS

Food irradiation is not a miracle food preservation method. Ionizing radiation exhibits both advantages and limitations, like any other method of food preservation. Food irradiation is neither as wonderful as some backers claim, nor as terrible as critics charge (24).

Urbain (25) reported the research completed since the 1940s demonstrates that treating foods with ionizing radiation (a) preserves foods to a varying extent, as determined by the treatment; food irradiation is particularly effective in controlling food-borne spoilage microorganisms; (b) decontaminates food of pathogenic bacteria, yeasts, molds, and insects; this decontamination can improve the hygienic quality of the foods and prevent potential health hazards; (c) controls maturation, senescence, and sprouting of fresh fruits and vegetables; (d) can alter chemical composition for quality improvement; (e) produces no toxic residues in foods; (f) retains nutritive value of foods; and (g) maintains sensory quality or results in generally better quality than that obtained from alternative procedures.

Among the advantages of food irradiation, one can mention the ability to replace chemical treatments that are increasingly coming under suspicion or are even being banned. The functional requirements of nitrites in cured meat products can be substantially reduced when the products are irradiated (26). Fumigation of food and food ingredients with various chemicals, such as ethylene dibromide, methyl bromide, and ethylene oxide, is either prohibited or being increasingly restricted in many advanced countries for health, environmental, or occupational safety reasons (12,13). Irradiation is an effective alternative to the chemical fumigants. More than a quarter of harvested food is lost because of various kinds of spoilage and waste. Pathogenic bacteria often contaminate spices and grains, but molds are more often the dominant microflora. Traditionally, fumigation with ethylene oxide was used to inactivate naturally occurring microflora on spices and grains. However, because of possible mutagenic and carcinogenic effects associated with the use of ethylene oxide, irradiation is now being used commercially as a sterilization treatment for spices, and demonstrates potential for application to disinfection of cereal grains (16). Irradiation potentially reduces the contact of many foods with chemicals designed to prevent the growth of infesting microorganisms. Many food scientists express more concern about fumigants than irradiation. Irradiation can replace, or drastically reduce, the use of food additives, fumigants, and food preservatives that pose potential hazards for the consumer as well as personnel in the food processing industry (7). Crop losses due to insect infestation and spoilage range from 10 to 25% in most subtropical countries, and up to 50% in some cases (26). One advantage of food irradiation over chemical fumigation is the application of irradiation to the packaged product, avoiding post-process contamination after irradiation treatment (6).

Gamma irradiation of peanuts effectively reduces the populations of natural mold contaminants and artificially inoculated *Aspergillus parasiticus* (27). However, at irradiation doses up to 15 kGy, complete elimination of the molds was not achieved. Under artificial abusive conditions, particularly under a high relative humidity, the molds in peanuts eventually grew, resulting in toxin production whenever water activity was not a limiting factor. Therefore, a combination of gamma irradiation to reduce the natural microbial population and environmental control practices such as proper drying, packaging, and storing at low relative humidity is recommended for preserving safe, high quality peanuts.

Narvaiz et al. (28) irradiated almonds and cashew nuts with doses between 1 and 4 kGy. The applied radiation doses reduced the initial mold and yeast load to acceptably low values that were unchanged for 6 months. No significant differences in the sensory quality of control and irradiated fruits were reported at the beginning of the storage period. After 6 months, a slight decrease in odor intensity was noticed in almonds irradiated with 1.5 and 2.0 kGy doses. The free fatty acids content and refractive index of the extracted oils were not affected by irradiation. However, increased lipid peroxidation was detected, but not to an extent noticeable organoleptically. Wilson-Kakashita et al. (29) compared the quality of English walnuts treated by gamma radiation at 5, 10, 15, and 20 kGy to propylene oxide (PO) treated and untreated walnuts. Gamma radiation at doses of 5 kGy or greater was more effective than PO in controlling mold. Irradiation of 10 kGy or greater was equal to or better than PO in controlling mesophylic aerobes. Gamma radiation and the PO treatment were equally effective in reducing coliforms to undetectable levels. Other than the peroxide value, there was no indication that gamma radiation induced rancidity in walnut lipids. Based on these results, gamma radiation at dose levels of 5–10 kGy is not only a viable alternative for PO but may be more effective than PO for controlling the microbial population of English walnuts.

An estimated 25% of world food crops is contaminated by mycotoxins. Crops most often contaminated include corn, peanuts, cottonseed, wheat, barley, and rice, with soy beans contaminated to a lesser extent. Little is known about the effects of irradiation on mycotoxin concentrations and protein quality in seeds. Hooshmand and Klopfenstein (30) reported that gamma irradiation doses of 5, 7.5, 10, or 20 kGy did not significantly reduce the aflatoxin B_1 concentration in soybeans at 9, 13, or 17% moisture, and had little effect on aflatoxin B_1 contamination of corn and wheat. Although moisture content is likely to be a major factor in determining the effectiveness of irradiation in destroying aflatoxins, no effect could be measured at the concentrations tested. At the maximum allowable irradiation dose for food (10 kGy), significant reductions in deoxynivalenol (DON) in soybeans (33%), zearalenone in corn (25%), and T-2 toxin in wheat (16%) were observed. Irradiation of soybeans reduced DON concentrations to meet current US Food and Drug Administration (FDA) advisory concentrations in feedstuffs. Reduction in the three toxins was smaller between 10 and 20 kGy of irradiation than between 0 and 10 kGy irradiation. Irradiation doses of 7.5 to 20 kGy destroyed 7, 13, and 7% lysine in corn, wheat, and soybeans, respectively. Methionine was significantly reduced in corn (17.5%) and wheat (17.9%). Phenylalanine and histidine were reduced to a lesser extent. Irradiation did not significantly change the concentrations of isoleucine, leucine, tyrosine, threonine, cysteine, valine, and arginine. No significant interaction on amino acid concentration occurred between irradiation and moisture effects. Essential amino acids in soybeans were more stable than those in other grains. Further studies of irradiation detoxification of grains may investigate the effects of using irradiation in combination with other "hurdles" such as H_2O_2, in order to reduce the radiation dose requirement and could result in safer and more nutritious food (30).

Lépine (31) considered the small number of studies dealing with the effects of irradiation on pesticides in food products and suggested it is difficult to draw conclusions about the extent of pesticide degradation, the nature of by-products, and the overall effects of irradiation on these compounds. Degradation of pesticides is generally greater in an irradiated aqueous solution than in aliphatic solvents or in food.

Irradiation can be applied to foods in small packages or in bulk, in a frozen state or at room temperature, with great versatility in applications. Prepackaged frozen foods can provide an oxygen-free atmosphere, thus avoiding oxidative degradation or heat denaturation (26). Foods do not warm appreciably during irradiation, allowing the use of irradiation on frozen foodstuffs. Irradiation is the only preservation method available for inactivating pathogenic microorganisms such as *Salmonella* in frozen foods (6,26).

One advantage of irradiation is that the treatment in many cases occurs after packaging, thus avoiding recontamination. Packaging materials used for irradiation must comply with regulations. Up-to-date packaging technology allows the use of combination processes, such as irradiation and modified atmosphere packaging. Modern resins are available to create multi-ply, coextruded films and thermoformed trays to meet the most stringent needs of food suppliers. The films provide moisture and gas barriers tailored to specific products for successful marketing. However, most of the packaging materials approved for food irradiation were developed before irradiation was available (32,33). One of the major potential applications of irradiation is the sterilization of containers in aseptic packaging (34). There is a need for more studies on available packaging materials. Packages will probably change to meet the specific needs of the irradiation process. Packaging must be resistant to irradiation with respect to their functional protective characteristics and not transmit toxic substances and undesirable odors or flavors to the food.

Thermal pasteurization of liquid food is a well established and widely accepted means of preservation, although not suited for solid foods such as meat, poultry, seafood, or dried ingredients of fresh foods whose raw characteristics must be maintained to meet market requirements (13). Alternative chemical preservation procedures for such foods have regulatory limitations and/or inherent public health problems due to toxic residues and environmental pollution. Much research data and many commercial experiences demonstrate that irradiation can play an important role in reducing some food-borne diseases.

Irradiation slows the rate of food deterioration, an advantage that consumers might eventually find appealing. This longer shelf life also enables farmers to keep more of their produce for a longer period of time. The use of irradiation to reduce viable populations of microorganisms in foods, with a consequent extension of time before incipient spoilage occurs, has been demonstrated by many researchers (16). The initial microbial population was effectively decreased by treating wrapped corn with 0.5 or 1.0 kGy irradiation (35). Narvaiz (36) extended the shelf life of mushrooms with a 3-kGy dose of gamma radiation. Weight loss, acidity, and compression and shear test texture analyses of mushrooms demonstrated that irradiation did not produce deleterious effects on the measured parameters for 17 days. Irradiation of minced fresh fish with 0.66 and 1.31 kGy extended the shelf life 6 and 13 days, respectively (37). Shelf-life extension of minced fish allows more time between mince and manufacture of finished products, like *surimi*, without freezing. Because of the destruction of the microorganisms that cause objectionable odors and flavors, a taste panel rated irradiated minced fish having rather high total aerobic plate counts as sensorially acceptable. This observation may indicate the desirability of having a safety margin in microbiological standards for irradiated fishery products. Low dose irradiation of 0.5–1.0 kGy extended the shelf life of fresh iced catfish fillets by 13 days (38). Paul et al. (39) indicate that low dose gamma irradiation reduces the initial microbial load in chunk or minced lamb and extends the shelf life at 0–3°C. Meat irradiated at 2.5 kGy possessed better color, odor, microbiological quality, and overall acceptability, with lower total volatile basic nitrogen values resulting in enhanced shelf life, compared to meat samples treated with 1.0 kGy. Improvement in the microbiological quality and shelf life of buffalo meat stored at ambient temperature by 2.5-kGy gamma irradiation was achieved by Naik et al. (40).

A number of food-borne parasitic diseases such as trichinosis, toxoplasmosis, taeniasis, and opisthorchiasis, prevalent in a number of developing countries, can also be controlled by low dose irradiation. In most cases, 1-kGy irradiation can render these parasites noninfective without causing notable changes in the physicochemical or sensory properties of the treated food (12,13).

Irradiation affects different foods in distinct ways. It breaks long molecules, such as cellulose, into shorter carbohydrates. Cellulose is the main constituent of plant cell walls; some fruits and vegetables become soft and lose some of their characteristic texture when irradiated. The irradiation of fats creates free radicals that oxidize fats and leads to rancidity. High dosages of irradiation can produce intense off-flavors resembling "burnt feathers" from oxidation or related off-odors. Irradiation can break down proteins and destroys a portion of some vitamins, particularly A, B, C, and E. But the low levels of irradiation recommended for use in food should ensure that deleterious processes do not happen at a higher level than in other methods of food processing, including cooking (7).

Amylose content of brown rice was not greatly affected by irradiation at 1–3 kGy, but rheological parameters decreased with increasing levels of radiation. As radiation increased, an increase in the yellow color of the rice was noted (41).

Mitchell et al. (42) reported that low-dose irradiation (< 0.6 kGy) did not appear to result in an increased juice yield from apples, grapes, and oranges. Irradiation treatment caused only nominal changes in fruit juice composition.

The effects of gamma irradiation (1–10 kGy) on selected functional properties were studied by Rao and Vakil (43) on four legumes: green gram, lentil, horsebean, and Bengal gram. Water absorption capacity of irradiated legumes increased, although pasting temperature was not appreciably changed. Irradiation with doses of 2.5–10.0 kGy of the legumes significantly reduced cooking time and the force required to compress cooked legumes.

Ceci et al. (44) concluded that the irradiation of garlic bulbs cv. Red did not adversely affect flavor. The flavor of preserved garlic is important becasue garlic is used principally as a condiment in the human diet.

Some disadvantages of irradiation processing turn out to be no more serious than the ones related to other, well-accepted preservation processes. It is said, for instance, that not all foods are suitable for irradiation treatment, but there is no preservation method that can be used on all foods (6). Some fruits may soften and discolor. Shell eggs cannot be irradiated to inactivate *Salmonella*, because the viscosity of egg whites is significantly affected at doses that effectively eliminates *Salmonella*, making the processed eggs unacceptable as table eggs (26). Milk acquires an unpleasant taste when irradiated because dairy products are particularly susceptible to oxidative flavor changes during treatment with ionizing radiation. Many food products (e.g., meat, chicken, fish) present threshold irradiation doses for organoleptic alterations (13). However, some of the organoleptic alterations can be offset if the food is irradiated in the frozen state. Certain protein foods are flavor-sensitive to irradiation and develop off-flavors. Irradiation, inhibits some biochemical processes such as enzyme activity in the food, delaying ripening and senescence, thus extending the period until the food product is ready for market (9). However, the maximum radiation dose that most fresh fruits and vegetables can tolerate without serious loss of firmness, ripening abnormalities, altered flavor, or increased susceptibility to mechanical injury is about 2.25 kGy (45). Doses of irradiation above 1.0 kGy may be detrimental to the quality of 'Climax' berries because of softening and loss of flavor and texture (46). Irradiated fruit of superior quality can be placed on the market when 'Rainier' cherries are treated with gibberillic acid before irradiation (47).

It is feared by some that treatment with ionizing radiation will convert a low quality or spoiled product into a seemingly high quality product by masking the off-flavors (48). In reality, this is not possible: irradiation can eliminate the spoilage organisms but cannot mask the off-flavor or odor of a spoiled product (26).

At the doses recommended for treating food, irradiation of certain foods, like other sub-sterilizing food processes, will not eliminate all microorganisms or their toxins. Low-dose irradiation will not destroy all bacterial spores (12). Thus, irradiated meat, chicken, and seafood will require appropriate temperature control similar to thermal pasteurized or controlled atmosphere storage of these products (e.g., to prevent microbial germination and toxin formation). The highly radiation resistant spore-forming species, such as *Clostridium botulinum*, survive the maximum approved dose, and this is often cited as a disadvantage of food irradiation. However, the same applies to many other preservation methods short of sterilization (6).

Vitamin losses are also often mentioned as a particular disadvantage of irradiation. In fact, the losses caused by irradiation processing are, in most cases, not of nutritional significance and are smaller than the ones caused by various heating processes (6). Ascorbic acid (vitamin C) is the most sensitive. Niacin, thiamine, riboflavin and beta carotene (provitamin A) are relatively stable (26).

A clear disadvantage of irradiation compared to thermal processing is the lack of enzyme inhibition in foods, even by high dose irradiation required for commercial sterilization. However, products that are shelf stable for years without refrigeration can be obtained when a sterilizing radiation dose is combined with other "hurdles," such as a mild blanching treatment to inactivate endogenous enzymes (6). The greatest disadvantage of food irradiation is its name (6). Ionizing radiation evokes unpleasant associations of radioactivity, nuclear threats, high technology, genetic mutation, and cancer.

Irradiation, like the other methods of food preservation, has a number of limitations. One way to overcome these limitations is to use irradiation in combination with other "hurdles," such as heating, modified atmospheres, vacuum packaging, or other methods of food preservation. Irradiation demonstrates its greatest potential in combination with other treatments (49). Successful combinations will enhance the preservative action and/or reduce the severity of other treatments. These combinations should provide improved product quality without reducing food safety. Combination treatments that show promise include irradiation and mild heat treatment; and modified atmosphere packaging; and curing salts. The use of combination processes often inhibits the development of undesirable sensory and chemical changes in foods (50), making food irradiation a more useful method of food preservation. Thakur and Arya (51) report gamma irradiation (10 kGy) produces beany and painty flavors in sweetened orange juice and mango pulp. Incorporation of 0.134% potassium sorbate before irradiation completely prevented the formation of irradiation-induced sensory changes and diminished the rate of browning and ascorbic acid losses during irradiation and 3 months' storage of orange juice and mango pulp.

Food irradiation is often compared to thermal pasteurization as a technology to assure the safety of foods. Irradiation is not a substitute for proper sanitation; but in combination with Good Manufacturing Practices, irradiation can reduce the risk of food-borne illnesses (16). Food irradiation is safe and its advantages outweigh its limitations (52). It is essential that interactions between the different variables involved in food irradiation be fully explored, especially with respect to organoleptic characteristics so that the boundary between advantage and disadvantage can be determined (19).

V. CHEMICAL, NUTRITIONAL, AND MICROBIOLOGICAL CHANGES IN IRRADIATED FOOD

A. Chemical Changes

Chemical changes will be induced in foods as a result of processing with irradiation. The objective of food irradiation is achieved as a result of these chemical changes. It is desirable to minimize the accompanying chemical changes in the food itself, not least because of the requirement for maintaining quality and avoiding undesirable odors and flavors. Ionizing radiation interacts with an irradiated material by transferring energy to electrons, which are thus raised to a higher-energy, or excited, state. If the transferred energy is large enough, the negatively charged electron can leave a molecule and form a positive ion. Irradiation induces about one ionization for every two excitations but because ionizations are about one thousand times as likely to induce chemical change, the biological effects caused by radiation may be entirely due to ionization. Ionizing radiation act quite differently from other forms of chemical or physical agents used to inactivate bacteria in that these other forms do not discriminate between molecules in an irradiated sample (53).

Many factors influence the type and extent of chemical change that may occur as a result of irradiation. The chemical complexity of many foods makes a precise prediction of the changes produced by irradiation almost as impossible as the prediction of changes produced by heat (54). However, studies on many individual food components, either in pure systems or food media, and on the combined effects of irradiation and other treatments have been made.

Nawar (55) reports that the effects of irradiation on the lipid fraction in foods are qualitatively similar to irradiation of natural fats or model systems analogous to fats. The products of lipid irradiation represent induced auto-oxidation products as well as non-oxidative radiolytic products. Moreover, the auto-oxidation products found are identical to those usually present in nonirradiated but oxidized fats. The extent of irradiation-induced oxidative changes depends on factors that are also associated with oxidation such as temperature, oxygen availability, fat composition, and pro-oxidants.

Denaturation of proteins is a result of irradiation, although a small fraction of protein molecules are denatured. The complexity of protein molecules themselves plus the presence of other components in food provide many loci for interaction with irradiation such that the products of radiolysis can be expected to be diverse and present in small amounts. The effects of irradiation on food proteins are reviewed by Swallow (56).

Irradiation of water gives rise to hydroxyl radicals, hydrated electrons, and other species. These, in turn, react with protein, producing protein free radicals. The radicals also react with other molecules in food. Some protein radicals are formed by direct action and via radiolysis of fat. Side chain radicals arise by oxidative and reductive attack on cysteine and cystine, and oxidative attack on methionine. Side chain radicals also arise by oxidative and reductive attack on histidine, tryptophan, tyrosine, and phenylalanine. Main chain radicals are formed by removal of the α-carbon hydrogen, and by reduction of the peptide carbonyl. Free radicals do not persist in food except in special circumstances, but the resulting chemical changes account for effects reported in proteins. The extent of radiolytic reactions is too small to cause appreciable changes in the amino acid content of food proteins, but significant color changes are sometimes reported and detectable volatiles are formed. In the case of meats and chicken, many volatiles, resulting from the indirect action of irradiation, are identified in parts per billion amounts (57). The amino acid composition of protein-containing foods, including meat, is not adversely affected by irradiation.

Although enzymes can be readily inactivated by irradiation in dilute aqueous solutions, enzymes in foods are relatively resistant to radiolysis (56). Destruction of many food enzymes requires five to ten times the irradiation dosage needed to inactivate microorganisms (58). Enzyme action may continue after microorganisms are inactivated unless a blanching treatment designed to inactivate enzymes preceded irradiation.

Irradiation of sugars in the pure state leads to marked degradation processes and the formation of radiolytic products. However, up to a dose of 1 kGy, the concentration of these products may be considered negligible. Moreover, the physical or chemical changes produced in sugars by an irradiation dose less than 1 kGy are small and less extensive than changes observed following thermal treatment (59). The influence of water on the radiolysis of pure sugars is complex. When sugars are irradiated in the solid state, water exerts a protective effect. The protective effect differs when irradiating sugars in solution. Because many foods rich in sugars also contain a large quantity of water, sugars in sweet foods undergo little change on irradiation. Further chemical studies are required on the radiolysis of sugars in mixtures or in foods, particularly on dry or dehydrated products.

Increasing dosages of gamma irradiation create increasing intensities of carbohydrate free radicals (60). Propagation of free radical reactions is intensified by a low moisture content. The carbohydrate free radicals initiate molecular changes in starches. Increasing the irradiation dosage promotes degradation of starch molecules, resulting in decreased viscosity and increased water solubility of the starch. The acidity of irradiated starch solutions increases with increasing irradiation dosages. Controlled irradiation of starchy foods can be used to make the foods safe for consumption. The observations of Sabularse et al. (61) indicate alterations of rice grain structure and composition due to gamma irradiation, and suggest that starch may be the major component of the rice kernel affected by irradiation.

Fatty acids are not affected by irradiation, except for palmitic acid, which exhibits a decrease, and oleic acid, which exhibits an increase in chicken carcasses treated with about 2.0 kGy gamma irradiation under commercial conditions (62).

Irradiation of potatoes increases discoloration and the concentration of phenols, and decreases the concentration of crude lipids and phospholipids (63). Irradiation of soybean seed with 7.5% moisture demonstrated little effect on the phospholipid composition and the thin layer chromatography pattern. On the other hand, gamma irradiation exposure at smaller irradiation doses and higher moistures resulted in severe damage to phospholipids and also affects their thin layer chromatography patterns, while significantly increasing inorganic phosphorus concentration and decreasing the phytate content of soybean seeds (64). Little information is available regarding the radiolytic products of phospholipids and phosphates of inositol. Research in this area is needed for a more complete understanding of the radiolysis of natural components such as phosphates of inositol and polar lipids. In general, irradiation-induced changes in complex foods are less extensive than the ones expected from irradiation studies carried out in pure components (65).

In addition to direct chemical changes, irradiation may alter the secondary and tertiary structures of proteins. Such changes may activate enzymes and cause physiological changes that may be detectable. The texture may be softened, the viscosity may be reduced, and the electrical conductivity may be increased (20).

The chemical approach is necessary to solve the key problems of detectors and irradiation-induced changes, and Johnston and Stevenson (66) discuss important current issues on the chemistry of food irradiation.

Changes of metabolism, chemical constituents, and organoleptic qualities induced by irradiation are discussed along with an evaluation of the shelf life of a wide variety of commodities in the reviews by Thomas (67–70).

B. Nutritional Changes

One of the more contentious issues is the nutritional adequacy of irradiated foods. Irradiation chemistry of many vitamins has been studied, but the relationship of the studies to the irradiation of foods is not clear (71). Although the irradiation stability (i.e., percentage retention) has been studied extensively, at least for some vitamins, relatively few studies include identification of vitamin degradation products formed due to irradiation. Vitamin E is the most irradiation-sensitive of the fat-soluble vitamins, and vitamin B_1 is the most radiolabile of the water-soluble vitamins. Many vitamins react differently toward heat and irradiation, but vitamins with little radiostability are susceptible to degradation by light, oxygen, or heat (72).

Carbohydrates, lipids, and proteins will not be noticeably depleted by irradiation, but minor components may be disproportionately depleted by susceptibility to attack by free radicals formed during irradiation. In practice, thiamine, ascorbic acid, and vitamins A and E are the most irradiation-sensitive vitamins (9,73). The United States Department of Defense irradiated 60,000 kg of bleached, enriched, hard wheat flour in 1967 and 1969. After 48 months of storage in multiwall paper bags or rectangular metal cans at 21°C, no differences in losses of thiamine, riboflavin, pyridoxine, and niacin were observed between irradiated flour (0.3–0.5 kGy) and unirradiated flour. Similarly, no differences in the content of the four vitamins were observed between breads baked with irradiated flour and unirradiated flour stored for 48 months (74). Mitchell et al. (75) reported that mangos and peppers can be irradiated at doses optimal to achieve disinfestation with significant retention of carotene.

Gamma irradiation of vacuum-packaged, ground fresh pork resulted in a dose-dependent first-order thiamine destruction rate. Thiamine losses (with respect to the nonirradiated sample) for raw pork irradiated at 0.57, 1.91, 3.76, 5.52, and 7.25 kGy were 7.7, 23.5, 38.1, 49.8, and 57.6%, respectively. Time of storage had little effect on the thiamine content of raw irradiated and nonirradiated pork (76). Fox et al. (77) studied the loss of thiamine and riboflavin due to gamma irradiation of beef, lamb, pork, and turkey. Thiamine losses averaged 11% per kGy and riboflavin losses 2.5% per kGy above 3 kGy dose. Loss of riboflavin was not different among meats. Any detriment from such slight losses would seem to be more than compensated by the advantage of controlling bacteriological contamination by irradiation processing. Commodities that are significant sources of ascorbic acid fortunately require low doses of irradiation for sprout inhibition or radurization. Consequently, most reported losses of vitamin C in vegetables and fruit are in the relatively low range of 0 to 20% (78). Abdellaoui et al. (79) demonstrated that irradiation at doses of 0.3 and 0.5 kGy, as an alternative to quarantine, preserved the organoleptic quality of clementines and the vitamin C content of the treated clementines was equivalent to that of control clementines. Gamma irradiation at 1 kGy reduces populations of aerobic mesophilic and lactic microflora in *pico de gallo*, a Mexican-style salsa (80). Sensory attributes and head space gases of the salsa are not affected. Ascorbic acid content declines 50% immediately after irradiation, but differences are slight between irradiated and nonirradiated salsa at the end of six weeks of storage at 2°C. Irradiation is a promising treatment for extending the shelf life of refrigerated pico de gallo.

The significance of vitamin retention in irradiated foods is widely debated. Proponents of irradiation argue that currently permitted irradiation doses present no nutritional problems. Vitamin retention following irradiation is comparable to other methods of food preservation and the impact on well-nourished populations will be insignificant. However, Bloomfield (73) suggests more research on the effects of commercial scale irradiation on retention of some nutrients such as vitamin A is needed.

C. Microbiological Changes

The number of food-borne bacteria recognized as human pathogens is increasing. The emergence of pathogens has increased interest in using irradiation as a preservation technique in the food industry (16). Tritsch (81) concludes that *Salmonella* and *Trichina* are inactivated by adequate cooking, and irradiation would not be necessary; but as Gould (82) points out, if the pathogens of concern did not enter the home or catering establishment in the first place, then the momentary lapses of hygiene that will always occur, at some frequency or other, would be of little consequence.

The incidence of food-borne diseases is increasing in many countries (6) and fortunately the most common and troublesome pathogenic bacteria are sensitive to irradiation and can be reliably inactivated by doses smaller than 10 kGy. The more resistant pathogens will be reduced in numbers, and the surviving flora is generally less resistant to other factors such as heat, pH changes, salt concentration and antibiotics; therefore, a combination of irradiation with other food preservation methods can be used to achieve the inactivation of pathogens. WHO estimates that up to 70% of diarrheal infections causing some 25% of deaths in developing countries are transmitted in food (83). Studies by the Centers for Disease Control and Prevention in the United States conclude that food-borne diseases caused by pathogenic bacteria such

as *Salmonella, Campylobacter, Escherichia coli, Vibrio, Trichinella,* tapeworms, and other parasites result in approximately 9000 deaths and from 6.5
to 81 million cases of diarrheal disease annually. Medical costs and losses of
productivity are also staggering. The enormous tragedy is that the losses are
preventable when current knowledge and proven technology are applied (8).

Ionizing radiation is lethal for bacteria (84). For the vast majority of
bacteria, the critical target for inactivation is the chromosome, a single, circular molecule of DNA containing several million base pairs. Most studies
indicate that damage to microbial DNA resulting in the loss of ability to reproduce is a primary cause of lethality, but damage to other sensitive and
critical molecules (e.g., in membranes) may also promote inactivation (85).
The proportion of a bacterial population that survives a select dose of irradiation will depend on the intrinsic sensitivity of the microorganism, its stage
during the growth cycle, the amount of irradiation damage inflicted, its potential for repair and the relative proportions of damage channeled into different repair pathways (84). Irradiation sensitivity of microorganisms
differs with species and even with strain, although the range of resistance
among strains of a single species is usually small enough to be negligible
(85). Gram-negative bacteria, including common food spoilage organisms,
and enteric species, including pathogens, are generally more sensitive to irradiation than are gram-positive bacteria. Irradiation resistance generally
follows the sequence (86):

gram negative < gram positive ≈ molds < spores ≈ yeasts < viruses

Bacterial endospores are more resistant to the action of ionizing radiation
than their corresponding vegetative cells by a factor of 5 to 15 (53). In general, the irradiation resistance of molds is equivalent to vegetative bacteria.
Yeasts are distinctly more resistant to irradiation than are molds, and as resistant as the more resistant bacteria. Viruses are even more irradiation resistant
than bacteria, so that irradiation treatments that destroy bacteria will not reliably inactivate viruses (85).

The bactericidal efficacy of a given dose of irradiation depends on the
following: kind and species of organism, number of organisms (or spores)
originally present, condition of the organisms, extracellular environmental
conditions such as pH, temperature, and chemical composition of the food,
presence or absence of oxygen, and physical state of the food during irradiation (16,22,58,87,88). An excellent review of information describing the behavior of pathogenic and spoilage microorganisms upon exposure to
irradiation, with a summary of irradiation D values of most microorganisms
of concern in foods is presented by Monk et al. (16). D is the radiation dose

required to eliminate 90% of a bacterial population (one logarithmic cycle reduction). Table 2 presents the susceptibility of various microorganisms to irradiation. The vast majority of the pathogenic bacteria have a D value less than 1 kGy, and for many spoilage bacteria the D value is lower than 10 kGy.

Gram-negative bacteria commonly involved in the spoilage of refrigerated fresh meats are sensitive to irradiation. Generally, gram-positive spoilage bacteria such as lactobacilli and lactococci are more resistant to gamma irradiation than are gram-negative bacteria. Irradiation resistance may result in a drastic change in the microbial flora in a food treated with low-dose irradiation. For this reason, the use of low-doses of irradiation that

TABLE 2   Irradiation D Values for Food-borne
Microorganisms

Microorganism	D (kGy)
Pathogenic bacteria	
Aeromonas hydrophila	0.04–3.40
Bacillus cereus (vegetative cells)	0.02–0.58
B. cereus (spores)	1.25–4.00
Campylobacter jejuni	0.08–0.32
Clostridium botulinum (spores)	0.41–3.20
C. perfringens (vegetative cells)	0.29–0.85
Escherichia coli	0.23–0.45
E. coli O157:H7	0.24–0.47
Listeria monocytogenes	0.25–0.77
Salmonella	0.37–0.80
Staphylococcus aureus	0.26–0.45
Yersinia enterocolitica	0.04–0.39
Vibrio	0.08–0.44
Spoilage bacteria	
Clostridium sporogenes	2.30–10.90
Micrococcus radiodurans	12.70–14.10
Moraxella phenylpyruvica	0.63–0.88
Pseudomonas putida	0.08–0.11
Sporolactobacillus inulinus (spores)	2.10–2.58
S. inulinus (vegetative cells)	0.35–0.53
Streptococcus faecalis	0.65–0.70
Viruses	2.02–8.10

D=radiation dose required to eliminate 90% of
microbial population (one log cycle reduction).
Source: From Refs 16 and 91.

would inactivate gram-negative bacteria but not lactic acid bacteria have potential application in fermented sausage production (13,16). Highly acceptable, microbiologically safe meat and meat products can be preserved by irradiation processing (89).

Spoilage of fish and other seafood by bacteria initially present in large populations occurs within a few days of harvest (16,37,38) even when storage is near 0°C. Contamination of seafood with pathogens such as *Salmonella* and *Vibrio* species is also a concern. *Vibrio* spp., especially *V. cholerae, V. parahaemolyticus,* and *V. vulnificus* are responsible for a number of disease outbreaks and deaths in recent years (13). Gamma irradiation is an effective method of inactivating microorganisms in seafood (12,13,16).

Of the well known food-borne pathogens, *Salmonella* is the most frequently encountered on poultry and meats. *Salmonella* is the most important causal agent of food-borne infection in most countries. Gram-negative bacterial pathogens, *Escherichia coli, Yersinia enterocolitica, Aeromonas hydrophila,* and *Campylobacter* species, exhibit relatively low resistance to irradiation (16,90). Of the gram-negative pathogens, *Salmonella* may be the most resistant. Hence, irradiation processes designed to eliminate *Salmonella* will also eliminate other gram-negative pathogenic bacteria. Published data on the radiosensitivities of *Salmonella, Campylobacter,* and *E. coli* O157:H7 are reviewed by Radomyski et al. (91).

Reduction factors of between 10^4 and 10^5 for *Salmonella* can be achieved by irradiating dehydrated eggs up to 3 kGy in the presence of air, and up to 5 kGy in the absence of air, without adversely affecting the organoleptic properties (92). Gamma irradiation of freshly processed chicken carcasses at 2.5 kGy delays microbial growth resulting in extended shelf life. Irradiation also reduces the initial loads of *Yersinia* and *Campylobacter* and the counts remain lower in irradiated poultry than in unirradiated through 18 days of storage at 4°C. *Salmonella* incidence can be completely eliminated by irradiation (93). Irradiation is an effective method for the control of *Escherichia coli* O157:H7 on foods(16,91).

A gamma irradiation (94) treatment of 6.5 kGy is more effective than ethylene oxide in reducing the mesophilic aerobic, coliform, sulfite-reducing anaerobic, yeast, mold, and *Salmonella* populations in paprika to permissible levels recommended by the International Commission on Microbiological Specification for Food.

An important cause of food intoxication throughout the world is *Staphylococcus aureus.* Although enterotoxigenic *S. aureus* may originate from foods of animal origin, contamination via humans may be more common in terms of number of cases of intoxication. Irradiation is an effective technique for inactivating *S. aureus* cells from meat products (16).

An increasing number of refrigerated foods is appearing in Japanese, European, and American markets. These include low-acid, refrigerated food products that represent specific challenges to food processors and distributors. Psychrotrophic pathogens such as *Yersinia enterocolitica, Aeromonas hydrophila,* and *Listeria monocytogenes* are food poisoning microorganisms that survive and grow in refrigerated foods at temperatures as low as 0°C (91). Because the psychrotrophic bacteria can grow in many refrigerated foods, other "hurdles" are often incorporated into refrigerated foods to limit and prevent growth of psychrotrophic pathogens during low-temperature storage. Irradiation is an efficient means to achieve significant reductions of psychrotrophic pathogens with minimal influence on other characteristics of the foods (16,91).

The yellowing of the rind in red and soft cheeses after irradiation (95), clearly detected only when compared to control cheeses, is a minor drawback compared with the ability to inactivate *L. monocytogenes*, even at a relatively high concentration (10^5 CFU/g), without inducing noticeable modifications to the taste, odor, or texture properties. A treatment dose of 2.6 kGy does not alter the organoleptic properties of Camembert cheeses, yet inactivates 10^4 *L. monocytogenes*/g (96). When 10^5 bacteria/g are inoculated, some bacteria are still viable after irradiation, and after 45 days. However, surviving bacteria often lose the ability to reproduce. The effects of combined electron beam irradiation and *sous vide* treatments on chicken breasts were studied by Shamsuzzaman et al. (97). *L. monocytogenes* was undetectable in chicken breasts treated with the combined treatment at 3.1 kGy, but survived the *sous vide* treatment without irradiation and reproduced during storage. Some loss of thiamine occurred with the combined treatments.

The effect of low-dose (0.1 kGy) irradiation on microflora, sensory characteristics, and development of oxidative rancidity of vacuum-packaged pork loins was investigated by Mattison et al. (98) after irradiation and during low temperature (4°C) storage up to 21 days. Irradiation reduced the numbers of mesophiles, psychrotrophs, anaerobic bacteria, and staphylococci with the effect on mesophiles and psychrotrophic spoilage organisms being the greatest. The effect of irradiation on sensory characteristics of pork loins was minimal, with no detectable differences between irradiated and nonirradiated pork after 14 days of storage. Irradiation of pork did not affect cooking loss or thiobarbituric acid (TBA) values.

A dose of 40 kGy irradiation adequately sterilized ice cream and frozen yogurt, but not mozzarella cheese (99). Resuscitation of the 40-kGy irradiated products by incubating them at two temperatures (7° and 35°C) aerobically and anaerobically for 8 weeks showed that no organisms survived in ice cream or yogurt, but *Staphylococcus*, *Streptococcus*, *Lactobacillus*, and *Bacillus* were detected in the irradiated cheese. The 12D for *B. cereus* preinoculated into cheese and ice cream was 43–50 kGy. This information should be considered if irradiated dairy products are used in low microbial diets for immunosuppressed patients. The gamma irradiation resistance of five enterotoxic and one emetic isolate of *B. cereus* vegetative cells and endospores was tested by Thayer and Boyd (100) in mechanically deboned chicken, ground turkey breast, ground beef round, ground pork loin, and beef gravy. The results indicated that irradiation of meat or poultry provides significant protection from vegetative cells but not from endospores of *B. cereus*.

Thermal processing of canned foods is based on the 12D concept, to ensure that the number of viable *Clostridium botulinum* spores is reduced from 10^6 to a risk of 10^{-6}. Because *C. botulinum* spores are irradiation and heat resistant, the 12D concept is applied to the process of radappertization designed to kill viable cells, including microbial spores of food spoilage and public health significance (16). Irradiation doses of 10 kGy are unlikely to inactivate spores of toxigenic bacteria including *C. botulinum* unless spores occur in small numbers in the food. High doses of irradiation are necessary to inactivate botulinum toxin in food. Ground pork meat was cured with the addition of $NaNO_2$, pasteurized, and irradiated with 0, 3, 6, 9, 10, 20, 30, 40, and 50 kGy; after irradiation, the samples were inoculated with a mixture of *Clostridium botulinum* type A, B, and E spores, incubated at 30°C for 28 days, and tested for the presence of botulinal toxin (101). Doses up to 9 kGy decreased $NaNO_2$ content with a corresponding decrease in antibotulinal activity. High irradiation doses resulted in a dose-dependent decrease in nitrite content with a corresponding loss of antibotulinal activity. After opening, or accidental perforation, of the package and contamination with *C. botulinum* spores, radappertized cured meats may be a better culture medium for this microorganism than cured meats preserved by heat treatment. Staphylococcal enterotoxin is more resistant to irradiation than botulinal toxin (16).

In general, bacteria become more resistant to ionizing radiation in the frozen state and also when dehydrated. In both cases, the contribution of indirect effects from the radiolysis of water is substantially reduced. Such effects need to be kept in mind, for example, if food is irradiated at very low temperatures to reduce the off-flavors caused by the oxidation of fats—because if the main consideration is the inactivation, or significant reduction in the numbers of contaminating bacteria, the irradiation dose would need to be increased to give the same level of microbial destruction (53). Proteolysis facilitates bacterial contamination of animal tissue. Alur et al. (102) studied the role of bacterial proteolysis in the spoilage of irradiated flesh foods. An appreciable delay in the spoilage of irradiated flesh foods can be attributed to reduced bacterial proteolysis.

Although irradiation reduces the mold population in foods, there is some uncertainty about the effects of irradiation on subsequent production of mycotoxins by surviving mycelia (16). Irradiation may result in subsequent enhancement, no effect, or a diminution of mycotoxin production. Reduction of competitive microflora allows the production of greater amounts of mycotoxins than would occur in the presence of effective competition. Surviving molds are expected to grow more rapidly in the absence of competitors and eventually dominate the microflora.

Radomyski et al. (91) suggest that in recommending the irradiation doses necessary to reduce or inactivate food-borne pathogens in foods, consideration of the nature of the food product, how it will be handled, its intended use, and other conditions such as temperature, pH, packaging atmosphere, water activity, and additives with antimicrobial activity have to be taken into account.

Fears are often expressed that microorganisms surviving an irradiation treatment may show enhanced pathogenicity or enhanced resistance to irradiation. Although the Joint Expert Committee has devoted much attention to the microbiological aspects of food irradiation, the FAO and WHO desired additional assurance that nothing had been overlooked in this area (6). At their request, the Board of the International Committee on Food Microbiology and Hygiene of the International Union of Microbiological Societies met in Copenhagen in 1982 to reconsider the evidence concerning microbiological safety of ionizing irradiation. The Board concluded that food irradiation is an important addition to the methods of control of food-borne pathogens and does not present any hazards to human health (103).

An international advisory group (104) reports no scientific evidence suggests irradiation results in increased induction of mutants with enhanced pathogenicity, virulence, or irradiation resistance. The advisory group adds that conventional processing techniques may also increase mutation rates, yet no evidence suggests increased pathogenicity or virulence of pathogenic organisms. The report concludes that even based on worst-case assumptions, induced radioactivity will be considerably smaller than the level occurring naturally in food. Irradiation in the commercially useful range is not expected to generate measurable additional radioactivity in foods. In 1993, the American Medical Association Council on Scientific Affairs affirmed food irradiation as a safe and effective process that increases the safety of food when applied according to governing regulations (8).

The wholesomeness, microbiological safety, nutritional adequacy, and lack of mutagenicity, teratogenicity, and toxicity of irradiated foods have been studied extensively. During many years of experience, the careful review of data and concerns raised reaffirms that irradiated foods are wholesome (2,78,105–108).

VI. SOME COMMERCIAL APPLICATIONS

Commercial food irradiation is currently being carried out by about 30 pilot and commercial plants in 25 countries, with additional irradiators planned or under construction in about ten other countries (9) The number of countries that use irradiation for processing food for commercial purposes has steadily increased (13) from 19 in 1987 to 34 in 1994. Food irradiation offers a strong potential to counter insect infestation, microbial spoilage, premature ripening, and physiological changes of food crops that result in postharvest food losses in developing countries (13).

Relatively low irradiation doses of 0.05–0.15 kGy are effective in inhibiting sprouting of stored bulbs and tubers such as potatoes (Fig. 2), onions,

FIGURE 2   Inhibition of sprouting in potato by irradiation. (Courtesy of Nordion International, Inc.)

and garlic, and are under investigation with other root crops (9). Wolters et al. (109) reported that 0.06 kGy irradiation almost completely prevented sprouting and improved the general appearance of onions. The storage losses were 50% for unirradiated and 20% for irradiated onions shipped from Argentina to the Netherlands. Sprout inhibition is attributed to DNA damage in the dormant cells at the tip of the sprout inhibiting cell division (9).

Absorbed irradiation doses in the range of 0.15–0.50 kGy are used to disinfest products such as grain, grain products, and fresh and dried fruit. The treatment relies on the fact that insects are killed by relatively low doses of 0.01–1 kGy ionizing radiation. No negative effects on the quality of the grains have been observed. Irradiation disinfestation is also applicable to barley, corn, rice, sorghum, legumes, corn flour, coffee, cocoa, and soy and to dry products such as dried fish, dried fruit, nuts, and tobacco; the irradiation dose required depends on the species of insect. Irradiation does not prevent subsequent reinfestation, so appropriate post-irradiation measures such as proper packaging or anoxic storage must be applied. Another important application of radiation disinfestation is the treatment of fresh tropical and subtropical fruit such as papaya (Fig. 3), mango, and citrus fruit. Many of these fruits are

FIGURE 3   Irradiated (right) and nonirradiated papayas (left). (Courtesy of Nordion International, Inc.)

infested by fruit flies and other insects. Pest-free countries impose strict quarantine controls to avoid importing the insects. Ethylene dibromide (EDB) fumigation is often used for disinfestation, but small quantities of potentially carcinogenic EDB are residual on the fruit, and thus its use is banned in many countries. Irradiation is one of the most effective alternatives available, and a minimum dose of 0.15 kGy was approved in the United States in January 1989 as a method for disinfesting papaya grown in Hawaii. Irradiation doses of 0.5–1 kGy delay maturation of fresh fruit and vegetables such as bananas, mango, oranges, papaya, and mushrooms (9). A hot benomyl dip, in combination with the irradiation of mangoes at doses between 0.07 and 0.6 kGy, preserves mango color and flesh texture with negligible phytotoxic effects (110).

The shelf life of foods such as strawberries, other fresh fruits and vegetables, fresh fish, and meat can be extended by an irradiation dose of 1.5–3.0 kGy, sufficient to reduce the population of most of the bacteria and molds responsible for spoilage. Strawberries represent an important application of such radurization because the microorganisms responsible for spoilage grow at low temperatures and cannot be controlled by refrigeration. Irradiation at a dose of 2.0–2.5 kGy retards spoilage, and the strawberries remain firm and fresh for extended periods (Fig. 4). Irradiation treatment of fresh fruit is limited to relatively small doses of less than 3 kGy because of softening and loss of quality at larger doses. Small-dose irradiation is leading to the development of combinations of processes for products such as citrus fruit, peaches, and cherries, combining irradiation with other "hurdles" such as heat or hot water treatments to limit the irradiation dose required for preservation (9).

A number of animal products are distributed under frozen conditions. Frozen distribution may be replaced by using a small-dose irradiation of 2–3 kGy together with chilling at 0–3°C to obtain sufficient shelf life for marketing (13). A study conducted by a large poultry processor in Brazil concluded that using such combined treatments instead of freezing resulted in a benefit of $0.18/kg poultry to the company, with sufficient shelf life to meet marketing requirements (12). Bacterial loads on chicken carcasses can be reduced by more than 99% with about 2.0-kGy gamma irradiation under commercial conditions without adversely affecting sensory or nutritional values (62). The good marketing experience of four stores selling irradiated poultry in the United States is described by Pszczola (111). In the case of poultry products, food irradiation offers a cost-effective method for assuring consumer protection against food-borne diseases, particularly salmonellosis and campylobacteriosis (17).

Whole and ground spices are contaminated with heat-resistant bacterial

FIGURE 4 Strawberries treated by irradiation. (Courtesy of Nordion International, Inc.)

spores and molds. Irradiation with 10 kGy holds promise for microbial inactivation without affecting the quality of commercially available, prepackaged Indian ground spices (112). The review of Farkas (4) includes a wealth of data on reducing microbial contamination and preventing insect damage through use of ionizing irradiation on dry food ingredients, herbs, and enzyme preparations.

A number of applications of ionizing irradiation in food processing leads to potentially valuable uses, including increasing the hydration rate of dehydrated vegetables, as in soup mixes; increasing the yield of juice from grapes without affecting the wine-making quality; increasing the rate of drying in fruits such as prunes; reducing the cooking time of dried beans; increasing the loaf size of bread made from flour used in formulas with small amounts of added sugars; reducing the amount of barley needed in beer production by increasing the yield of the malted grain; and reducing the flatulence-producing propensity of beans (3).

Suppliers of equipment and processing services are slow to develop efficient technology for irradiating both medical and food products in the same facility. Irradiators designed specifically for food processing—requiring a low dose—limit competition in the more lucrative and established medical market (113).

Compared to the total food market, the niche for irradiated foods is still small, even in countries where permission for several irradiated food items is approved (6). Regulatory and safety aspects of the irradiation process, equipment and cost factors, and necessary consumer acceptance contribute to hesitant commercialization of food irradiation. Successful implementation of a new technology depends upon the availability of a proper infrastructure within a given country. In general, food irradiation processing requires the same infrastructure as other physical processes such as canning, freezing, drying. Compared with the energy required for canning, refrigeration, and/or frozen storage, the energy used for irradiation preservation of food is small (13).

Understandably, any food treatment adds costs to the product. Like other physical processes, irradiation requires large capital costs and critical minimum capacity and product volumes for economic operation (5). However, unlike other physical processes, irradiation operating costs are small, especially with regard to energy requirements. In technologically advanced nations, as well as in many developing countries, the infrastructure required for setting up irradiation processing plants exists (12,13). However, there are other factors (114) that are at least equally important in determining the rate of commercialization of irradiated foods, including the food processors themselves. Hall (115) discusses the necessary decisions surrounding successful food processor investment in irradiation processing.

Transfer of technology from government and academia to the food industry in France resulted in successful commercialization of irradiation, and as Boisseau (116) points out, there are four basic steps necessary for successful marketing of irradiated foods: (a) research must define conditions of irradiation application, (b) legislation must specify conditions of application, (c) consumers must accept the irradiated food product, and (d) appropriate processing capacity must exist. Camembert cheese and mechanically deboned poultry meat are successful examples of commercialization of irradiated foods in France.

One example of food irradiation's potential to prevent parasite infections is the irradiation of *nham* in Thailand. *Nham* is a fermented pork and rice sausage that can be contaminated by *Salmonella*, *Shigella*, and other pathogenic bacteria, as well as by parasites such as *Taenia*, *Trichinella*, and *Entamoeba*. *Nham* is very popular as a raw product because cooking changes the flavor unacceptably to many Thais. The Thai government sponsored a long-term study into the safety and technical feasibility of irradiation processing of *nham*. As a result of this study, *nham* irradiated with 2 kGy minimum is in the marketplace and accepted by consumers, although priced higher than its unirradiated counterpart. Labeling informs the consumer that *nham* is irradiated (6,17).

Similarly, consumers in Bangladesh accepted irradiated dried fish instead of fish fumigated with various pesticides, including DDT (13). Irradiated dried fish has been marketed with success since a semi-commercial scale food irradiator came into operation at Chittagong in early 1994. Because irradiated frog legs meet the strict microbiological specifications in France, such frog legs have been routinely marketed at the retail level for the past 10 years.

VII. CONSUMER ACCEPTANCE OF IRRADIATED FOOD

Consumer requirements constantly change—and with respect to irradiated foods, these requirements cover a wide range. However, the main issues are safety, wholesomeness, the consumer right to know, possible abuses of irradiation, and the lack of methods for detection of irradiated foods. Food irradiation is increasingly being recognized as an effective method for ensuring the hygienic quality of food. However, although a new process may be introduced because of technical advantages, the question that still remains is whether such a new process will be generally accepted by the consumer.

Attitudes toward food irradiation were studied by Defesche (117) and Cramwinckel and van Mazijk-Bokslag (118) in the Netherlands, and by Wiese (119), Bruhn et al. (120), and Bruhn and Schutz (121) in the United States. The authors state that consumer fear of food irradiation must be taken seriously. However, it is uncertain if consumer concerns are against the goal of making food more safe through irradiation. Public knowledge of food processing methods in general and food irradiation in particular is limited. Although accurate science-based information about food irradiation is now reaching consumers in the United States, there is still a long way to go, and the level of public knowledge in other countries is extremely limited (13,122). Many consumers do not understand why food should be irradiated while the same foods (that are not irradiated) are available in the marketplace (13).

The controversy over food irradiation started in the early 1980s when Australia, Canada, the United Kingdom, and the United States sought public comments on the process of introducing national regulations for food irradiation. Numerous self-appointed consumer advocacy groups opposing the introduction of food irradiation emerged (12,13). Consumer advocacy groups opposed to irradiated food exerted substantial political pressure in some countries and prevented the granting of any permission for food irradiation. In other countries, such as the United Kingdom, these groups delayed regulations for many years and have made certain that the regulations require licensing with a long and complicated list of conditions, provisions, and requirements unnecessary for other food processing methods. Faced with such complications and threats of boycotts and picketing, the food industry in some countries lacks enthusiasm for promoting irradiation. Because of the efforts of the advocacy groups, irradiation is a political and psychological issue (6,123,124).

Consumer attitudes and concerns about food irradiation changed after education with factual information and opinions (125). In 1989, about 85% of UK consumers said they would not eat irradiated food (126). Recent surveys demonstrate that only a small minority would "never" buy irradiated food (24).

Contrary to public perception, results of more recent market tests and the commercial sale of irradiated food in 40 market trials in 20 countries were positive in favor of irradiated food. The opinion of consumers about irradiated food is quite different when they are given the opportunity to select and purchase the food (13). In many cases, consumers preferred irradiated food because of product quality. None of these studies provided any evidence to indicate that consumers would reject irradiated food (16,122). Results of a new study (127) suggest that consumers are less concerned about irradiation than about food additives, pesticide residues, animal drug residues, growth hormones, and bacteria. The risk to workers and environmental issues are among the top concerns regarding irradiation, and about one-third of consumers believe that irradiated food is radioactive. Bruhn (128) states that the majority of consumers will respond positively to irradiated foods when the advantages of irradiation are explained and when food safety, nutritional value, worker safety, and environmental concerns are addressed. An investment in consumer education is required to open the market for irradiated foods, but the rewards in food safety, food quality, and environmental safety are substantial.

The Advisory Committee on Irradiated and Novel Foods (129) advised that ionizing irradiation can be used safely to improve the shelf life of food. The Committee's response to 6000 letters received from members of the public and 150 letters received from organizations was that there is no evidence to support many of the concerns expressed in the letters: the formation of toxic radiolytic products, the possibility of free radicals being ingested, or the loss of nutritional value.

Diehl (130) reported that public knowledge about food irradiation is minimal, so unwarranted conclusions, misunderstandings, and half-truths are mixed in many books and articles written by those opposed to food irradiation. Most of the detractors of irradiation exaggerate the facts in an attempt to gain supporters (131,132). Consumer advocacy groups express concern about gamma irradiation facilities, which require special safety precautions and strictly controlled operating procedures in handling the radiation sources, and the environmental issues regarding nuclear waste. Regulation and inspection have to be the basic requirements for every food processing procedure, and not only for irradiation.

The widespread educational campaigns demanded by the WHO and articles for the general public like the ones by Ghosh (7), Castleman (24), Gunther (133), and Willman (15) are largely needed. It is essential that creative, technical minds come together to develop an organized plan for negating bad publicity because presentations and literature that deal only with the technical aspects of irradiation will not clear the concerns. The general public must be educated concerning the scientific rationale that supports the use of irradiation in foods and leads to current regulatory approval (124). Educational efforts should include analyses of the comparative benefits and risks of irradiation processing relative to other established food processing techniques. The educational task must be a cooperative effort between the government and the food industry to provide a broad basis of support and a strong credibility. A closer relationship with the media must be developed (123). In the late 1980s and early 1990s, universities, professional societies, and industry groups included irradiation in public information programs (13,122). National regulatory bodies throughout the world, United Nations agencies, and the food industry are taking an approach based on scientific information, extensive experience, and a genuine regard for the needs and concerns of the general public (134). Bruhn (128) recommends several science-based educational materials for the lay audience. Audiovisual educational programs have been effective in increasing both consumer knowledge and a positive attitude toward food irradiation (135).

Each modern food processing advance—pasteurization, canning, and freezing—produced criticism. Food irradiation is no different. Leaders in the health and food-related professions must dispel the myths and explain the facts (8). As pointed out by Schweigert, (136) food science and nutrition educators face the challenge of providing appropriate information to the public today, so that when irradiated foods become available for consumption, meaningful choices can be made by the consumer based on factual information. Consumer acceptance of food irradiation, like that of any new technology, will depend on the perceived risks and benefits (137). Consequently, promotion of the acceptability of food irradiation among consumers requires better understanding of consumer concerns and their relationship to the wider social and cultural values associated with food and the provision of information that addresses these concerns.

VIII. IRRADIATION AND THE INTERNATIONAL FOOD TRADE

Food plays an important role in international trade and the national economies of many nations of the world. Food trade is most important in developing countries where economies are still largely agricultural (138). As the world takes steps toward becoming a global economy, the serious issue of food safety increases. Whether it is a question of pesticide residues, the introduction of unlabeled, untested genetically engineered foods, or food-borne diseases, food safety is of paramount importance in trading countries.

Irradiation could potentially increase the food supply by reducing spoilage and increase trading opportunities by improving shelf life and hygienic quality (83). Irradiation is facilitating trade by replacing the use of ethylene dibromide, methyl bromide, and ethylene oxide fumigants widely used to overcome trade barriers in place to protect against insect infestation and microbial contamination of imported foods (12,13).

Although the irradiation of foodstuffs is approved in many countries as an important method for preventing food spoilage and food-borne diseases, worldwide acceptance and trade in irradiated foods is hindered by the lack of an appropriate mean of control.

A. Identification of Irradiated Foods

Detecting irradiated foodstuffs as such—to inform consumers and control domestic and international trade in irradiated food—is essential. The ability to reliably distinguish between irradiated and unirradiated foods and ingredients is fundamental to reassuring the public that consumer rights are protected (66). This is not a trivial problem, because the effects produced by irradiation are often small and may be similar to changes produced by other means of food preservation (20). The first step is to determine the "fingerprints" produced by irradiation, and then to consider possible processes that can give similar "fingerprints" in the food. At present there is no universal method for identifying irradiated foodstuffs, and different kinds of radiation effect (chemical, physical, and biological) are under investigation for this purpose (139,140). Among the radiation effects are changes in histological and morphological characteristics of food as a result of irradiation, changes in microflora, physical properties (electrical, impedance, viscosity, wettability), chemical changes in food components (carbohydrates, lipids, proteins, nucleic acids, vitamins), including the formation of volatile products, and the formation of free radicals stabilized in solid food components such as bones, shells, seeds, and achenes (9). It is unlikely that a single test applicable to every food will emerge. Glidewell et al. (141) concluded that most likely there will be a battery of methods, some suitable for rapid screening and others, probably used in combination, capable of delivering more information in return for a greater investment of time, money, and expertise. Broadly, whatever the analyte, two types of tests can be distinguished: first, a precise reference method that can be undertaken in a specialized laboratory, preferably leading to an estimation of the original radiation dose; and second, an inexpensive rapid screening method that can be used on the spot to identify suspect food samples (142).

Although much progress is being made, it is unlikely that rapid, routine analytical methods suitable for use in the enforcement of food irradiation legislation will be available in the near future for all foods (104). The compounds produced by irradiation of food packaging materials may potentially serve as indicators that the package was irradiated (143). The Codex Alimentarius Committee was originally pessimistic about the prospect of developing tests, but the progress made in recent years toward the development of methods for the identification of irradiated foods is significant and more rapid than anticipated. Useful detection methods are developed for specific foods. Research is continuing to extend the applications of these detection technologies to a wider range of foods.

Detection methods used include electron spin resonance (ESR), viscosity, thermoluminescence, lyoluminiscence, conductivity, chemical analysis of volatiles, microflora, and DNA molecular composition. The combined application of selected methods provides a more informative picture of irradiation-induced processes (144). The different methods for detecting irradiated foods were reviewed by a Working Group set up by the WHO (145), Glidewell et al. (141), and Rahman et al. (139,140).

Differences between flours obtained from different varieties of cereals and potatoes, as well as between unirradiated flour and irradiated flour, can be detected by applying small-angle x-ray scattering (146). Irradiation-induced free radicals in food containing bone or shells can be detected by ESR. Irradiated herbs and spices can be identified by thermoluminescence (147–151). Katusin-Razem et al. (92) explored the potential of the irradiation-induced oxidation of lipids in spray-dried eggs with the formation of hydroperoxides as a suitable chain reaction for identification. The study of Pfordt and von Grabowski (152) makes evident the detectability of irradiated eggs as an ingredient of bakery products. Nawar et al. (153) studied radiolytic products of lipids as markers for the detection of irradiated meats. For 81 meat products, the technique gave 100% correct determinations of irradiation. Stevenson et al. (149), and Stevenson (151,154) report that 2-alkylcyclobutanones are potential postirradiation markers for minced chicken meat and other animal products such as pork, lamb, beef, and liquid whole eggs. The ESR signals from frog legs, fish bones, and fruit seeds appear less stable but can often be used (20). The results of Tabner and Tabner (155) confirm the usefulness of ESR spectroscopy in determining the irradiation history of citrus fruits. Three international trials on the use of ESR to detect irradiated meats containing bone were sponsored as part of a coordinated program by the IAEA (156). The results were 100% successful in identifying both irradiated and unirradiated bones. The findings obtained by Onori and Pantaloni (157) suggest ESR can be used for qualitative and quantitative analysis of irradiated fresh eggs. Thermoluminescence was used in an interlaboratory study (158) to detect irradiation treatments of spices, herbs, and spice-herb mixtures in the dose range used for the reduction of microbial counts. Correct identification of irradiated or nonirradiated treatment was achieved in 99.1% of 317 spices and herbs.

The results in the detection of irradiated foods are very encouraging, and in the majority of cases it has been possible to identify foods treated with irradiation at doses well below those likely to be used commercially. However, further collaborative testing and international agreement on suitable methods are needed to increase international trade of irradiated foods. The similarity of the endogenous and irradiation-induced ESR signals restrict the practical utility of the ESR technique for the detection of irradiation treatment in mango fruits (159). Deighton et al. (160) conclude that chemical treatment of cellulosic material involving strong alkali during purification results in subtle structural alterations, and as a result sensitivity is adversely affected, with potentially serious implications for both quantitative and qualitative detection of irradiated cellulose and lignocellulosic material.

Although it is ideal to have a method that is specific for irradiation, there is interest in the use of rapid screening methods that give a preliminary indication that irradiation may have taken place; the suspicion can be confirmed by more definitive methods (154). Hydrogen is a promising marker for distinguishing between irradiated and unirradiated frozen food. Hydrogen could provide a rapid screening test, complementing other chemical, physical, and microbiological procedures (142). Positive detection of hydrogen provides conclusive evidence of irradiation. However, no definite conclusions can be drawn from a failure to detect hydrogen. A microbiological screening method for the detection of irradiation of frozen poultry meat was developed by Wirtanen et al. (161) on the basis of the combined use of total cell count by the direct epifluorescent filter technique and viable cell count by the aerobic plate count. However, other food processing methods give similar results, so this microbiological test is not specific for irradiation and the results need to be confirmed using another technique. Because most foods contain DNA, a detection method based on irradiation-induced changes in DNA may potentially have a wide application. As mentioned by Deeble et al. (162), the detection of irradiation-induced changes in DNA has great potential as the basis of an assay for irradiated food. A membrane-based DNA hybridization technique was used by Rowe and Towner (163) in a model system to examine the effect of irradiation on the detection of bacterial contamination in foods. Although hybridization signals were reduced compared with otherwise identical unirradiated foods, artificial contamination populations in excess of 10^5 CFU per test were distinguished in 12 of the 13 foods examined following the irradiation process.

Currently, the European Committee for Standardization is preparing standard methods for identification of irradiated foods. The methods being considered are ESR for foods containing bone and/or cellulose; thermoluminescence for spices and herbs; and the hydrocarbon and cyclobutanone methods for food containing fat (151).

Further research is needed to determine if these techniques can be used to estimate irradiation doses and whether the markers survive processing and storage conditions. Many irradiated foods are not amenable to the methods discussed, thus there is a need to extend the range of available methods.

B. Legislation

To promote worldwide introduction of food irradiation, it is necessary to develop national and international legislation and regulatory procedures to enhance confidence among trading nations that foods irradiated in one country and exported are irradiated with acceptable standards of wholesomeness, dosage of irradiation, and hygienic practice.

The recommendations in 1980 of the Joint Expert Committee and in 1983 of the Codex Alimentarius created a positive impact on national regulations. The Codex General Standard for Irradiated Foods serves as a model for individual countries; incorporating its provisions in national legislation will protect consumers and facilitate international trade (138).

Variation in the regulations among countries restricts trade of irradiated products. Some countries follow the recommendation of the World Health

Organization in setting the maximum dosage at 10 kGy, and others restrict irradiation dosage. For example, poultry can be irradiated at maximum dosages of 3 kGy in the United States and the Netherlands, 5 kGy in France, and 7 kGy in Bangladesh, Brazil, Chile, Israel, South Africa, Syria, Thailand, and the United Kingdom. Verification that a food processor is irradiating products within regulation limits is very difficult. Control of the irradiation process is similar to control in sterilizing canned products. With a canned product, it is also difficult to determine if it has been heated to a high enough temperature for ample time to meet the canning regulations. Times and temperatures must be recorded and inspections of records are part of the control process. Likewise, the irradiation dosage is determined by irradiation-sensitive materials called dosimeters. Some of the ways in which these measurements can be achieved are published by Sharpe (164). Dosimeters establish irradiation processes and assure absorbed doses. Records of dosimeter measurements are part of the control process. While regulation of the canning process is developed to a point where confidence in the system provides for confident international trade of canned products, regulation of irradiated products is not harmonized and will need to be more uniform to successfully incorporate irradiated products into international trade.

Labeling is an important issue related to harmonization of irradiated products. Some countries require irradiated foods to be labeled with the green radura symbol (Fig. 5) and words such as "irradiated, treated with irradiation, radura, protected by ionization, and treated by irradiation [Fig. 4]"; other nations demand just the radura symbol and no descriptive words; and no special label requirements are asked in other countries(15). The FDA approved irradiation for wheat and wheat powder, white potatoes, spices and dry vegetable seasonings, dry or dehydrated enzyme preparations and aromatic vegetable substances, pork, fresh fruits, and poultry (165). The FDA approved food irradiation under provisions of the Food Additives Amendment of the Federal Food, Drug, and Cosmetic Act that requires that an irradiated food not be sold in the United States unless the Department of Health and Human Services finds that the food is safe and issues a regulation specifying safe conditions of irradiation. Thus, food irradiation in the United States is considered a food additive and not a process, and the irradiated food must be labeled as such. Pauli and Tarantino (166) briefly outline the information necessary to issue an authorizing regulation, describe the conditions under which food may currently be irradiated, and discuss the basis for current regulations. Thus, consumers can be optimistic of the future of international trade in irradiated foods based on the developments mentioned.

FIGURE 5   The international symbol for irradiated foods.

The 1988 International Conference on the Acceptance, Control of, and Trade in Irradiated Food outlined steps to facilitate wider acceptance and control of international trade in irradiated foods (167). The international regulatory status was reviewed recently by Derr (33). An effective quality control program is the primary control method for irradiation processing of food. The program must be fully compatible with Good Manufacturing Practices or other similar sanitation practices and should be based on the Hazard Analysis and Critical Control Point (HACCP) system.

Considerable information regarding harmonization of regulations, codes of good irradiation practices, inventories of authorized food irradiation facilities, and guidelines to overcome certain trade restrictions in irradiated foods were developed by the International Consultative Group on Food Irradiation since 1984 (17). Considering the current and expected level of commercialization, the resolution of trade issues is extremely important, especially because the most likely product groups for irradiation processing are major export commodities. The positive conclusion in 1993 of the Uruguay Round of the GATT Multilateral Trade Negotiations, especially the adoption of an agreement on the Application of Sanitary and Phytosanitary Measures, has added further incentive to international trade in irradiated foods (13).

IX. INTERNATIONAL COOPERATION

Before introducing new food processing technologies, adequate and reliable evidence must provide sufficient assurance that the processes not only produce the desired results in food, but also do not have any unacceptable toxicological, nutritional, or microbiological effects (168). Several international organizations play an important role in the gathering of evidence at the international level for food irradiation. The organizations include the International Project in the Field of Food Irradiation (IFIP), World Health Organization (WHO), Food and Agriculture Organization (FAO), and International Atomic Energy Agency (IAEA). FAO, IAEA, and WHO recognized the potential role of food irradiation in increasing food availability and improving food safety since the late 1950s. Cooperation first started in 1961 when the three organizations convened an Expert Panel on Wholesomeness of Irradiated Foods (13). In 1976, the Joint FAO/IAEA/WHO Expert Committee (JECFI) on the Wholesomeness of Irradiated Foods recommended unconditional acceptance for human consumption of irradiated potatoes, wheat, strawberries, papaya, and chicken, and provisional acceptance of irradiated rice, onions, fresh cod, and redfish. The Joint Committee also recognized food irradiation as a physical process comparable to heating and freezing. In 1980, the JECFI concluded that irradiation of foods with an absorbed dose of up to 10 kGy causes no toxicological hazard and introduces no special nutritional or microbiological problems. These recommendations and conclusions were adopted in 1983 by the Codex Alimentarius Commission and incorporated into a Codex General Standard for Irradiated Foods. The basis for worldwide legal acceptance of irradiated foods now existed. The success of international cooperation in establishing the wholesomeness of irradiated food (up to a maximum dose of 10 kGy, which covers most applications of food irradiation) led the Board of Management of IFIP to recommend that FAO, IAEA, and WHO consider another form of international cooperation, to harmonize regulations in different countries and to facilitate commercial application of the technology for the benefit of mankind—and thus the International Consultative Group on Food Irradiation (ICGFI) was established in early 1983. Cooperation with the Codex Alimentarius Commission is continuing through the activities of the ICGFI (13).

FAO, IAEA, OECD, and a total of 24 countries cooperated in IFIP between 1970 and 1981 to study the wholesomeness of irradiated food and to supply data for evaluation by the JECFI. WHO participated in IFIP on an advisory basis. IFIP played a crucial role in developing, coordinating, and compiling toxicological, nutritional, and microbiological data related to the wholesomeness of irradiated food (13). Largely through the efforts of IFIP, the 1980 JECFI was able to conclude that irradiation of foods with an absorbed dose of up to 10 kGy causes no toxicological hazard and introduces no special nutritional or microbiological problems.

Many national governments that previously had not granted any permissions for irradiated foods did so after 1983. Few governments accepted the Joint Expert Committee verdict outright as a basis for national regulations on irradiated foods. The governments appointed their own expert committees that again went through the scientific evidence and verified the conclusions of the Geneva meeting of 1980. Still, health authorities over the world hesitated to approve the marketing of irradiated foods.

After another extensive review of the existing data, WHO released an updated policy statement on September 23, 1992.

> Irradiated food produced under established Good Manufacturing Practices is to be considered safe and nutritionally adequate because the process of irradiation will not introduce: a) changes in the composition of the food which, from a toxicological point of view, would impose an adverse effect on human health; b) changes in the microflora of the food which would increase the microbiological risk to the consumer; and c) nutrient losses in the composition of the food which, from a nutritional point of view, would impose an adverse effect on the nutritional status of individuals or populations.

In 1994, WHO (104) came out with a detailed endorsement of ionizing radiation. The report says, "Whereas pasteurization was introduced mainly to interrupt the transmission of bovine tuberculosis and brucellosis, the most important public health application of food irradiation is to destroy or reduce the ubiquitous and largely unavoidable pathogens that contaminate raw foods, especially those of animal origin."

The leading organization supporting food irradiation is the ICGFI in Vienna, Austria. ICGFI tracks global food irradiation, provides processor and consumer information, leads training programs, and assists nations in developing capabilities (15). Considerable information regarding harmonization of regulations, codes of good irradiation practice, inventories of authorized food irradiation facilities, and guidelines to overcome certain trade restrictions in irradiated foods were developed by this group. Such information is essential for arriving at legislative standards and bilateral and international trade agreements related to irradiated food (16). The membership of ICGFI more than doubled—from 15 governments in 1984 to 40 in 1994, the majority of which are developing countries. The governments provide cash or in-kind support to the activities of ICGFI (13). The mandate of the ICGFI, initially planned for 5 years, was extended twice, in 1989 and 1994.

Successful transfer of food irradiation was attained in Asian countries under the Asian Regional Cooperative Project on Food Irradiation (RPFI), operated by the FAO/IAEA. Technology transfer activities in food irradiation are also being conducted in the other regions, including the FAO/IAEA Technical Cooperation Program on Food Irradiation in Latin America (LAFIP), and developing countries in Europe and the Middle East, Europe, and Africa. In addition to the efforts of FAO and IAEA, global research and training in food irradiation are supported by the International Facility for Food Irradiation Technology (IFFIT) based in the Netherlands (138). FAO, IAEA, and the Ministry of Agriculture and Fisheries of the Netherlands cooperated in establishing IFFIT at the Institute of Atomic Research in Agriculture, Wageningen, the Netherlands, between 1978 and 1990. Trained manpower on food irradiation for developing countries is established mainly through IFFIT (13).

X. ADDITIONAL REMARKS

Of the food processing techniques, food irradiation has undergone the most prolonged and intensive analyses. Independent committees of scientists agree that irradiation, if applied correctly, is safe. Food irradiation is increasingly recognized as an effective method for reducing postharvest food losses, ensuring the hygienic quality of food, and facilitating wide trade of selected foods. More than 40 years of research on the efficacy of irradiation as a food preservation technique resulted in a copious number of scientifically sound publications. Irradiation has a useful contribution to make, but it is not the total solution: irradiation will neither replace Good Manufacturing Practices nor provide the sole answer to food-borne illnesses.

The use of irradiation will expand globally. Continual reduction and banning of fumigants will lead to the expansion of disinfecting food with irradiation. More countries are clearing products to be irradiated, and some of these nations are building irradiation facilities. As regulatory agencies similar to the FDA approve certain products, countries that do not have the resources to develop an extensive regulatory agency will also clear those products. Well defined standards for irradiated foods at the national and international level will be written. Standardized methods of detecting irradiated foods and irradiation dosages will be forthcoming. The central role of consumer acceptability in the market penetration of new technologies has to be realized, and a wide range of issues relating to the irradiation technology itself and the food system into which it is going to be introduced has to be addressed. As irradiation is utilized to preserve more food products, more consumers will come in contact with them. As consumers become more knowledgeable about irradiation and the safety and wholesomeness of irradiated foods their concerns will diminish.

Irradiation treatments are designed to reduce or inactivate microorganisms that may be present in foods. The combined effect of irradiation and other "hurdles" or modes of preservation can be used to ensure microbial safety and also prevent the development of undesirable sensory and chemical changes in some foods. Modified atmosphere packaging, vacuum packaging, barrier films, subjecting foods to high or low temperatures before, during, or after irradiation treatment, and the use of chemical preservatives work well in conjunction with irradiation to extend shelf life. As pointed out by Karel (34), the future of food technology will be to combine various methods of food preservation. Ultimately, the real gain from food irradiation may come when it is used in combination with other preservation techniques, using the "hurdle" effect concept described in the last chapter of this book. Irradiation will undoubtedly become an acceptable method of food preservation. The recent books of Diehl (169) and Murano (165) confirm this opinion, placing the entire subject of food irradiation into proper perspective.

REFERENCES

1. E. S. Josephson and M. S. Peterson, *Preservation of Food by Ionizing Radiation*, CRC Press, Boca Raton, Florida, 1982.
2. E. Wierbicki, E. S. Josephson, A. Brynjolfsson, J. M. Coon, R. B. Maxcy, C. Merritt, I. A. Taub, M. Doyle, G.W. Froning, J. L. Goodenough, G. D. Griffin, A. A. Kader, W. J. Lipton, A. McIntosh, S. O. Nelson, C. S. Perry, H. J. Reitz, M. G. Simic, J. S. Sivinski, J. W. Slavin, D. W. Smith, E. W. Tilton, and W. M. Urbain, *Ionizing Energy in Food Processing and Pest Control: I. Wholesomeness of Food Treated With Ionizing Energy*, Report 109. Council for Agricultural Science and Technology, Ames, IA, 1986.
3. E. Wierbicki, E. S. Josephson, A. Brynjolfsson, J. M. Coon, R. B. Maxcy, C. Merritt, I. A. Taub, M. Doyle, G. W. Froning, J. L. Goodenough, G. D. Griffin, A. A. Kader, W. J. Lipton, A. McIntosh, S. O. Nelson, C. S. Perry, H. J. Reitz, M. G. Simic, J. S. Sivinski, J. W. Slavin, D. W. Smith, E. W. Tilton, and W. M. Urbain, *Ionizing Energy in Food Processing and Pest Control: II. Applications*, Report 115. Council for Agricultural Science and Technology, Ames, IA, 1989.
4. J. Farkas, *Irradiation of Dry Food Ingredients*, CRC Press, Boca Raton, Florida, 1988.
5. W. M. Urbain, *Food Irradiation*, Academic Press, New York, 1986.
6. J. F. Diehl, Will irradiation enhance or reduce food safety. *Food Policy* 18(2): 143-151 (1993).

7. P. Ghosh, A taste for irradiated food. *New Scientist* 123: 40-41 (1989).
8. P. R. Lee, Irradiation to prevent foodborne illness. *JAMA* 272(4): 261 (1994).
9. R. J. Woods, Food irradiation. *Endeavour* 18(3): 104-108 (1994).
10. Anonymous, Perspective on food irradiation. *Food Technol.* 41(2): 100–101 (1987).
11. S. Hadlington, Food irradiation legislation due despite public suspicion. *Nature* 328: 751 (1987).
12. P. Loaharanu, Status and prospects of food irradiation. *Food Technol.* 48(5): 124-131 (1994).
13. P. Loaharanu, Food irradiation: Current status and future prospects, *New Methods of Food Preservation* (G. W. Gould, ed.), Blackie Academic & Professional, New York, 1995, pp. 90-111.
14. Anonymous, *Food Irradiation*, U.S. Army brochure, 1994.
15. M. L. Willman, As the world learns. *The National Provisioner* 209(1): 44-51 (1995).
16. J. D. Monk, L. R. Beuchat, and M. P. Doyle, Irradiation inactivation of foodborne Microorganisms. *J. Food Prot.* 58(2): 197-208 (1995).
17. P. Loaharanu, Cost/benefit aspects of food irradiation. *Food Technol.* 48(1): 104-108 (1994).
18. W. M. Urbain, Irradiated foods: a giant step beyond Appert. *Nutrition Today* 19(4): 2-7 (1984).
19. C. H. McMurray, Food irradiation: the challenge, *Food Irradiation and the Chemist* (D. E. Johnston and M.H. Stevenson, eds.), Royal Society of Chemistry, Cambridge, United Kingdom, 1990, pp. 1-12.
20. A. Brynjolfsson, Future radiation sources and identification of irradiated foods. *Food Technol.* 43(7): 84-89, 97 (1989).
21. T. Hayashi, Comparative effectiveness of gamma-rays and electron beams in food irradiation, *Food Irradiation* (S. Thorne, ed.), Elsevier Applied Science, New York, 1991, pp. 169-206.
22. J. M. Jay, *Modern Food Microbiology*, 3rd ed., Van Nostrand Reinhold Co., New York, 1986.
23. B. Hoekstra and G. H. Pauli, Questioning the safety of food irradiation (letter to the editor). *Public Health Reports* 108(3): 402 (1993).
24. M. Castleman, Radiant grub, *Sierra* 78: 27-28 (1993).
25. W. M. Urbain, Food irradiation: the past fifty years as prologue to tomorrow. *Food Technol.* 43(7): 76, 92 (1989).
26. M. R. Sahasrabudhe, Food irradiation: current status, concerns, limitations and future prospects. *Journal of the Canadian Dietetic Association* 51(2):329–334 (1990).
27. R. Y. Y. Chiou, C. M. Lin, and S. L. Shyu, Property characterization of peanut kernels subjected to gamma irradiation and its effect on the outgrowth and aflatoxin production by Aspergillus parasiticus. *J. Food Sci.* 55(1): 210-213, 217 (1990).

28. P. Narvaiz, G. Lescano, and E. Kairiyama, Irradiation of almond and cashew nuts. *Lebensm.-Wiss. u.-Technol.* 25: 232-235 (1992).

29. G. Wilson-Kakashita, D. L. Gerdes, and W. R. Hall, The effect of gamma irradiation on the quality of English walnuts (Juglans regia). *Lebensm.-Wiss. u.-Technol.* 28: 17-20 (1995).

30. H. Hooshmand and C. F. Klopfenstein, Effects of gamma irradiation on mycotoxin disappearance and amino acid contents of corn, wheat, and soybeans with different moisture contents. *Plant Foods for Human Nutrition* 47: 227-238 (1995).

31. F. L. Lépine, Effects of ionizing radiation on pesticides in a food irradiation perspective: a bibliographic review. *J. Agric. Food Chem.* 39(12): 2112-2118 (1991).

32. D. Kilcast, Irradiation on packaged food. *Food Irradiation and the Chemist* (D. E. Johnston and M. H. Stevenson, eds.), Royal Society of Chemistry, Cambridge, United Kingdom, 1990, pp. 140-152.

33. D. D. Derr, International regulatory status and harmonization of food irradiation. *Journal of Food Protection* 56(10): 882-886, 892 (1993).

34. M. Karel, The future of irradiation applications on Earth and in space. *Food Technol.* 43(7): 95-97 (1989).

35. T. Deak, E. K. Heaton, Y. C. Hung, and L. R. Beuchat, Extending the shelf life of fresh sweet corn by shrink-wrapping, refrigeration, and irradiation. *J. Food Sci.* 52(6): 1625-1631 (1987).

36. P. Narvaiz, Some physicochemical measurements on mushrooms (Agaricus campestris) irradiated to extend shelf life. *Lebensm.-Wiss. u.-Technol.* 27: 7-10 (1994).

37. H. A. Dymsza, C. M. Lee, L. O. Saibu, J. Haun, G. J. Silverman, and E. S. Josephson, Gamma irradiation effects on shelf life and gel forming properties of washed red hake (Urophycis chuss) fish mince. *J. Food Sci.* 55(6): 1745-1746, 1748 (1990).

38. L. A. Przybylski, M. W. Finerty, R. M. Grodner, and D. L. Gerdes, Effect of shelf life of iced fresh channel catfish fillets using modified atmospheric packaging and low dose irradiation. *J. Food Sci.* 54(2): 269-273 (1989).

39. P. Paul, V. Venugopal, and P. M. Nair, Shelf life Enhancement of lamb meat under refrigeration by gamma irradiation. *J. Food Sci.* 55(3): 865-866 (1990).

40. G. N. Naik, P. Pushpa, S. P. Chawla, A. T. Sherikar, and P. M. Nair, Improvement in microbiological quality and shelf life of buffalo meat at ambient temperature by gamma irradiation. *J. Food Safety* 13: 177-183 (1993).

41. V. C. Sabularse, J. A. Liuzzo, R. M. Rao, and R. M. Grodner, Physicochemical characteristics of brown rice as influenced by gamma irradiation. *J. Food Sci.* 57(1): 143-145 (1992).

42. G. E. Mitchell, A. R. Isaacs, D. J. Williams, R. L. McLauchlan, S. M. Nottingham, and K. Hammerton, Low dose irradiation influence on yield and quality of fruit juice. *J. Food Sci.* 56(6): 1628-1631 (1991).

43. V. S. Rao and U. K. Vakil, Effects of gamma-radiation on cooking quality and sensory attributes of four legumes. *J. Food Sci.* 50(2): 372-375, 378 (1985).

44. L. N. Ceci, O. A. Curzio, and A. B. Pomilio, Effects of irradiation and storage on the flavor of garlic bulbs cv "Red". *J. Food Sci.* 56(1): 44-46 (1991).

45. A. A. Kader, Potential applications of ionizing radiation in postharvest handling of fresh fruits and vegetables. *Food Technol.* 40(6): 117-121 (1986).

46. W. R. Miller, R. E. McDonald, and T. G. McCollum, Quality of 'Climax' blueberries after low dosage electron beam irradiation. *J. Food Quality* 17: 71-79 (1994).

47. S. R. Drake, H. R. Moffitt, and D. E. Eakin, Low dose irradiation of 'Rainier' sweet cherries as a quarantine treatment. *Journal of Food Processing and Preservation* 18: 473-481 (1994).

48. I. Anderson, Hot food paper left out in the cold. *New Scientist* 139:6 (1993).

49. M. F. Patterson, The potential for food irradiation. *Letters in Applied Microbiology* 11(2): 55-61(1990).

50. B. R. Thakur, and R. K. Singh, Combination processes in food irradiation. *Trends in Food Science and Technology* 6: 7-11 (1995).

51. B. R. Thakur, and S. S. Arya. Effect of sorbic acid on irradiation-induced sensory and chemical changes in sweetened orange juice and mango pulp. *Int. J. Food Sci. Technol.* 28: 371-376 (1993).

52. J. I. Ahmad, Food irradiation—Facts and fiction. *Nutrition and Food Science* 3: 29-31 (1995).

53. B. E. B. Moseley, Ionizing radiation: Action and repair, *Mechanisms of Action of Food Preservation Procedures* (G. W. Gould, ed.), Elsevier Applied Science, New York, 1989, pp. 43-70.

54. A. J. Swallow, Chemical effects of irradiation, *Radiation Chemistry of Major Food Components* (P. S. Elias and A. J. Cohen, eds.), Elsevier Scientific Publishing, New York, 1977, pp. 5-20.

55. W. W. Nawar, Radiation chemistry of lipids, *Radiation Chemistry of Major Food Components* (P. S. Elias and A. J. Cohen, eds.), Elsevier Scientific Publishing, New York, 1977, pp. 21-62.

56. A. J. Swallow, Effects of irradiation on food proteins, *Developments in Food Proteins—7* (B. J. F. Hudson, ed.), Elsevier Applied Science, New York 1991, pp. 195-231.

57. W. M. Urbain, Radiation chemistry of proteins. *Radiation Chemistry of Major Food Components* (P. S. Elias and A. J. Cohen, eds.), Elsevier Scientific Publishing, New York, 1977, pp. 63-130.

58. W. C. Frazier and D. C. Westhoff, *Food Microbiology*, 4th ed., McGraw-Hill Book Co., New York, 1988.

59. J. F. Dauphin and L. R. Saint-Lebe, Radiation chemistry of lipids, *Radiation Chemistry of Major Food Components* (P. S. Elias and A. J. Cohen, eds.), Elsevier Scientific Publishing, New York, 1977, pp. 131-186.

60. A. S. Sokhey and M. A. Hanna, Properties of irradiated starches *Food Structure* 12: 397-410 (1993).

61. V. C. Sabularse, J. A. Liuzzo, R. M. Rao, and R. M. Grodner, Cooking quality of brown rice as influenced by gamma irradiation, variety and storage. *J. Food Sci.* 56(1): 96-98, 108 (1991).

62. S. R. Katta, D. R. Rao, G. R. Sunki, and C. B. Chawan, Effect of gamma irradiation on whole chicken carcasses on bacterial loads and fatty acids. *J. Food Sci.* 56(2): 371-372 (1991).

63. N.I. Mondy and B. Gosselin, Effect of irradiation on discoloration, phenols and lipids of potatoes. *J. Food Sci.* 54(4): 982-984 (1989).

64. Y. S. Hafez, A. I. Mohammed, P. A. Perera, G. Singh, and A. S. Hussein, Effects of microwave and gamma irradiation on phytate and phospholipid contents of soybean (Glycine max. L.). *J. Food Sci.* 54(4): 958-962 (1989).

65. Y. S. Hafez, A. I. Mohammed, G. Singh, and F. M. Hewedy, Effects of gamma irradiation on proteins and fatty acids of soybean. *J. Food Sci.* 50(5): 1271-1274 (1985).

66. D. E. Johnston and M. H. Stevenson, *Food Irradiation and the Chemist*, Royal Society of Chemistry, Cambridge, United Kingdom, 1990.

67. P. Thomas, Radiation preservation of foods of plant origin. II. onions and other bulb crops. *CRC Critical Reviews in Food Science and Nutrition* 21(2): 95-136 (1984).

68. P. Thomas, Radiation preservation of foods of plant origin. III. Tropical fruits: bananas, mangoes, and papayas. *CRC Critical Reviews in Food Science and Nutrition* 23(2): 147-205 (1986).

69. P. Thomas, Radiation preservation of foods of plant origin. IV. Subtropical fruits: citrus, grapes, and avocados. *CRC Critical Reviews in Food Science and Nutrition* 24(1): 53-89 (1986).

70. P. Thomas, Radiation preservation of foods of plant origin. VI. Mushrooms, tomatoes, minor fruits and vegetables, dried fruits, and nuts. *CRC Critical Reviews in Food Science and Nutrition* 26(4): 313-358 (1988).

71. A. Rogan and G. Glaros, Food irradiation: the process and Implications for Dietitians *J. American Dietetic Assocation* 88(7): 833-838 (1988).

72. P. P. Tobback, Radiation chemistry of lipids, *Radiation Chemistry of Major Food Components* (P. S. Elias and A. J. Cohen, eds.), Elsevier Scientific Publishing, New York, 1977, pp.187-220.

73. L. Bloomfield, Food irradiation and vitamin A deficiency. *Food Policy* 18(1): 64-72 (1993).

74. E. S. Josephson, Food irradiation: a Canadian folly (book review). *Food Technol.* 47(1): 139 (1993).

75. G. E. Mitchell, R. L. McLauchlan, T. R. Beattie, C. Banos, and A. G. Gillen, Effect of gamma irradiation on the carotene content of mangos and red capsicums. *J. Food Sci.* 55(4): 1185-1186 (1990).

76. R. K. Jenkins, D. W. Thayer, and T. J. Hansen, Effect of low-dose irradiation

and post-irradiation cooking and storage on the thiamin content of fresh pork. *J. Food Sci.* 54(6): 1461-1465 (1989).

77. J. B. Fox, L. Lakritz, J. Hampson, R. Richardson, K. Ward, and D. W. Thayer, gamma irradiation effects on thiamin and riboflavin in beef, lamb, pork, and turkey. *J. Food Sci.* 60(3): 596-598, 603 (1995).

78. J. H. Skala, E. L. McGown, and P. P. Waring, Wholesomeness of irradiated Foods. *J. Food Prot.* 50(2): 150-160 (1987).

79. S. Abdellaoui, M. Lacroix, M. Jobin, C. Boubekri, and M. Gagnon, Effets de l'irradiation gamma avec et sans traitement à l'eau chaude sur les propriétés physicochimiqes, la teneur en vitamine C et les qualités organoleptiques des clémentines. *Sci. Aliments* 15(3): 217-235 (1995).

80. L. R. Howard, G. H. Miller, and A. B. Wagner, Microbiological, chemical, and sensory changes in irradiated pico de gallo. *J. Food Sci.* 60(3): 461-464 (1995).

81. G. L. Tritsch, Food irradiation (letter to the editor). *Nutrition Reviews* 50(10): 311 (1992).

82. G. W. Gould, ed., *New Methods of Food Preservation*, Blackie Academic & Professional, New York, 1995.

83. A. McGregor, Food irradiation endorsed, *Lancet* 344: 1149 (1994).

84. B. E. B. Moseley, Radiation, micro-organisms, and radiation resistance, *Food Irradiation and the Chemist* (D. E. Johnston and M. H. Stevenson, eds.), Royal Society of Chemistry, Cambridge, United Kingdom, 1990, pp. 97-108.

85. M. Ingram and T. A. Roberts, Ionizing irradiation, *Microbial Ecology of Foods. Volume I* (edited by ICMSF), Academic Press, Inc., New York 1980, pp. 46-69.

86. M. R. Adams and M. O. Moss, *Food Microbiology*, The Royal Society of Chemistry, Cambridge, United Kingdom, 1995.

87. L. R. Beuchat, Gamma irradiation: a tool for eliminating pathogens in red meats and poultry. Presented at Annual Midwest Food Processing Conference, LaCrosse, Wis., October 3-4, 1994.

88. P. Dion, R. Charbonneau, and C. Thibault, Effect of ionizing dose rate on the radioresistance of some food pathogenic bacteria. *Can. J. Microbiol.* 40: 369-374 (1994).

89. D. W. Thayer, R. V. Lachica, C. N. Huhtanen, and E. Wierbicki, Use of irradiation to ensure the microbiological safety of processed meats. *Food Technol.* 40(4): 159 (1986).

90. M. F. Patterson, Sensitivity of Campylobacter spp. to irradiation in poultry meat. *Letters in Applied Microbiology* 20: 338-340 (1995).

91. T. Radomyski, E. A. Murano, D. G. Olson, and P. S. Murano, Elimination of pathogens of significance in food by low-dose irradiation: a review. *Journal of Food Protection* 57(1): 73-86 (1994).

92. B. Katusin-Razem, B. Mihaljevic, and D. Razem, Lipid test, *Nature* 345: 584 (1990).

93. P. O. Lamuka, G. R. Sunki, C. B. Chawan, D. R. Rao, and L. A. Shackelford,

Bacteriological quality of freshly processed broiler chickens as affected by carcass pretreatment and gamma irradiation. *J. Food Sci.* 57(2): 330-332 (1992).

94. S. Llorente Franco, J. L. Gimenez, F. Martinez Sanchez, and F. Romojaro, Effectiveness of ethylene oxide and gamma irradiation on the microbiological population of three types of paprika. *J. Food Sci.* 51(6): 1571-1572,1574 (1986).

95. S. Ennahar, F. Kunts, A. Strasser, M. Bergaentzle, C. Hasselman, and V. Stahl, Elimination of *Listeria monocytogenes* in soft and red smear cheeses by irradiation with low energy electrons. *Int. J. Food Sci. Technol.* 29: 395-403 (1994).

96. D. L. Bougle and V. Stahl, Survival of *Listeria monocytogenes* after irradiation treatment of Camembert cheeses made from raw milk, *J. Food Prot.* 57(9): 811-813 (1994).

97. K. Shamsuzzaman, L. Lucht, and N. Chuaqui-Offermanns, Effects of combined electron-beam irradiation and *sous-vide* treatments on microbiological and other qualities of chicken breast meat. *J. Food Prot.* 58(5): 497-501 (1995).

98. M. L. Mattison, A. A. Kraft, D. G. Olson, H. W. Walker, R. E. Rust, and D. B. James, Effect of low dose irradiation of pork loins on the microflora, sensory characteristics and fat stability. *J. Food Sci.* 51(2): 284-287 (1986).

99. A. E. Hashisaka, J. R. Matches, Y. Batters, F. P. Hungate, and F. M. Dong, Effects of gamma irradiation at -78°C on microbial populations in dairy products. *J. Food Sci.* 55(5): 1284-1289 (1990).

100. D. W. Thayer and G. Boyd, Control of enterotoxic Bacillus cereus on poultry or red meats and in beef gravy by gamma irradiation. *J. Food Prot.* 57(9): 758-764 (1994).

101. J. Szczawinski, M. Szczawinska, and M. Szulc, Effect of irradiation on antibotulinal efficacy of nitrite. *J. Food Sci.* 54(5): 1313-1317 (1989).

102. M. D. Alur, S. B. Warrier, S. N. Doke, and P. M. Nair, Role of bacterial proteolysis in the spoilage of irradiated flesh foods. *Journal of Food Biochemistry* 17: 419-432 (1994).

103. Codex Alimentarius Commission, *The Microbiological Safety of Irradiated Food*, Report CX/FH 83/9, Rome, 1983.

104. WHO, *Safety and Nutritional Adequacy of Irradiatied Food*, World Health Organization, 1994.

105. Anonymous, *Report on the Safety and Wholesomeness of Irradiated Foods*, Advisory Committee on Irradiated and Novel Foods. Department of Health and Social Services, Northern Ireland, 1985.

106. P. S. Elias, New concepts for assessing the wholesomeness of irradiated foods. *Food Technol.* 43(7): 81-83 (1989).

107. N. Craft, Food irradiation is safe, says WHO. *BMJ* 309(6962): 1105 (1994).

108. D. W. Thayer, Wholesomeness of irradiated foods, *Food Technol.* 48(5): 132-135 (1994).

109. T. C. Wolters, D. I. Langerak, O. A. Curzio, and C. A. Croci, Irradiation effect

on and onion keeping-quality after seashipment from Argentina to the Netherlands. *J. Food Sci.* 55(4): 1181-1182 (1990).

110. A. J. Jessup, C. J. Rigney, and P. A. Wills, Effect of gamma irradiation combined with hot dipping on quality of "Kensington Pride" mangoes. *J. Food Sci.* 53(5): 1486-1489 (1988).

111. D. Pszczola, Irradiated poultry makes U.S. debut in Midwest and Florida markets, *Food Technol.* 47(11): 89-96 (1993).

112. M. A. Munasiri, M. N. Parte, A. S. Ghanekar, A. Sharma, S. R. Padwal-Desal, and G. B. Nadkarni, Sterilization of ground prepacked Indian spices by gamma irradiation. *J. Food Sci.* 52(3): 823-824, 826 (1987).

113. Anonymous, New irradiator sterilizes both food and medical products. *Chilton's Food Engineering* 65(1): 44-46 (1993).

114. J. G. Leemhorst, Commercial food irradiation in practice. *Food Irradiation and the Chemist* (D. E. Johnston and M. H. Stevenson, eds.), Royal Society of Chemistry, Cambridge, United Kingdom, 1990, pp. 153-167.

115. R. L. Hall, Commercialization of the food irradiation process *Food Technol.* 43(7): 90-92 (1989).

116. P. Boisseau, Irradiation and the food industry in France. *Food Technol.* 48(5): 138-140 (1994).

117. F. Defesche, Consumer attitudes towards irradiation of food, *Marketing and Consumer Acceptance of Irradiated Foods* (edited by Joint FAO/IAEA) International Atomic Energy Agency, Vienna, 1983, p.47.

118. A. B. Cramwinckel and D. M. van Mazijk-Bokslag, Dutch consumer attitudes toward food irradiation. *Food Technol.* 43(4): 104, 109-110 (1989).

119. Wiese Research Associates, *Consumer Reaction to the Irradiation Concept*, Omaha, Nebraska, 1984.

120. C. M. Bruhn, M. G. Schutz, and R. Somer, Attitude change toward food irradiation among conventional and alternative consumers, *Food Technol.* 40(1): 86-91 (1986).

121. C. M. Bruhn and H. G. Schutz, Consumer awareness and outlook for acceptance of food irradiation. *Food Technol.* 43(7): 93-94, 97 (1989).

122. C. M. Bruhn, Consumer attitudes and market response to irradiated food. *J. Food Prot.* 58(2): 175-181 (1995).

123. D. Pszczola, Food Irradiation: countering the tactics and claims of opponents. *Food Technol.* 44(6): 92-97 (1990).

124. M. C. Lagunas-Solar, Radiation processing of foods: an overview of scientific principles and current status. *J. Food Prot.* 58(2): 186-192 (1995).

125. V. A. McCracken and P. A. Dickrell, *Food irradiation: exploring consumer attitudes*, Research Bulletin XB1014, College of Agriculture and Home Economics Research Center, Washington State University, Pullman, 1989.

126. Anonymous, Irradiation essential to make food safe. *New Scientist* 122: 17 (1989).

127. A. V. A. Resureccion, F. C. F. Galvez, S. M. Fletcher, and S. K. Misra, Con-

sumer attitudes toward irradiated food: results of a new study, *J. Food Prot.* 58(2): 193-196 (1995).

128. C. M. Bruhn, Strategies for communicating the facts on food irradiation to consumers. *J. Food Prot.* 58(1): 213-216 (1995).
129. Anonymous, Not irradiation yet. *Nature* 331: 469 (1988).
130. J. F. Diehl, Thought for food. *Nature* 344: 207 (1990).
131. R. Piccioni, Food irradiation:contaminating our food. *The Ecologist* 18(2): 48-55 (1988).
132. D. Murray, Food irradiation: dangerous, ineffective and unnecessary. *Search* 23(4): 113(1992).
133. J. A. Gunther, The food zappers. *Popular Science* 244(1): 72-86 (1994).
134. WHO, *Food Irradiation. A Technique for Preserving and Improving the Safety of Food*, World Health Organization, Geneva.1988.
135. A. J. Pohlman, O. B. Wood, and A. C. Mason, Influence of audiovisuals and food samples on consumer acceptance of food irradiation. *Food Technol.* 48(12): 46-49 (1994).
136. B. S. Schweigert, Food irradiation: What is it? Where is it now? Where is it going? *Nutrition Today* 22: 13-19 (1987).
137. S. Henson, Demand-side constraints on the introduction of new food technologies: the case of food irradiation. *Food Policy* 20(2): 111-127 (1995).
138. P. Loaharanu, International trade in irradiated foods: regional status and outlook, *Food Technol.* 43(7): 77-80 (1989).
139. R. Rahman, A. K. M. M. Haqsue, and S. Sumar, Chemical and biological methods for the identification of irradiated foodstuffs. *Nutrition and Food Science* 1: 4-11 (1995).
140. R. Rahman, A. K. M. M. Haque, and S. Sumar, Physical methods for the identification of irradiated foodstuffs. *Nutrition and Food Science* 2: 36-41 (1995).
141. S. M. Glidewell, N. Deighton, B. A. Goodman, and J. R. Hillman, Detection of irradiated food: a review. *J. Sci. Food Agric.* 61: 281-300 (1993).
142. C. H. S. Hitchcock, Determination of hydrogen in irradiated frozen chicken, *J. Sci. Food Agric.* 68: 319-323 (1995).
143. D. W. Allen, A. Crowson, D. A. Leathard, and C. Smith, The effects of ionising radiation on additives present in food-contact polymers. *Food Irradiation and the Chemist* (D. E. Johnston and M. H. Stevenson, eds.), Royal Society of Chemistry, Cambridge, United Kingdom, 1990, pp. 124-139.
144. J. Kispéter, L. I. Horváth, K. Bajúsz-Kabók, L. I. Kiss, and P. Záhonyi-Racs, Identification of radiation treatment of mineral-enriched milk protein concentrate by complex test protocols: a Comparison of luminescence, electron spin resonance and rheological investigations. *Food Structure* 12: 379-384 (1993).
145. K. W. Bögl, D. F. Regulla, and M. J. Suess, Health impact, identification, and dosimetry of irradiated foods (1988) *Future Radiation Sources and Identification of Irradiated Foods* (A. Brynjolfsson, ed.). *Food Technol.* 43(7): 84-89, 97 (1989).

146. K. Ciesla, T. Zóltowski, and R. Diduszko, Physico-chemical changes occurring in gamma irradiated flours studied by small-angle x-ray scattering. *Food Structure* 12: 175-180 (1993).

147. D. C. W. Sanderson, Luminescence detection of irradiated foods, *Food Irradiation and the Chemist* (D. E. Johnston and M. H. Stevenson, eds.), Royal Society of Chemistry, Cambridge, United Kingdom, 1990, pp. 25-56.

148. M. H. Stevenson and R. Gray, Can ESR spectroscopy be used to detect irradiated food, *Food Irradiation and the Chemist* (D. E. Johnston, and M. H. Stevenson, eds.), Royal Society of Chemistry, Cambridge, United Kingdom, 1990, pp. 80-96.

149. M. H. Stevenson, A. V. J. Crone, and J. T. G. Hamilton, Irradiation detection. *Nature* 344: 202 (1990).

150. G. A. Schreiber, B. Ziegelmann, G. Quitzsch, N. Helle, and K. W. Bögl, Luminescence techniques to identify the treatment of foods by ionizing radiation. *Food Structure* 12: 385-396 (1993).

151. M. H. Stevenson, Identification of irradiated foods. *Food Technol.* 48(5): 141-144 (1994).

152. J. Pfordt and H. U. von Grabowski, Backwaren aus Bestrahlten und Unbestrahlten Eiern—Nachweis der Bestrahlung bei Einem Weiterverarbeiteten Lebensmittel *Z. Lebensm. Unters Forsch* 200: 198-202 (1995).

153. W. W. Nawar, Z. R. Zhu, and Y.J . Yoo, Radiolytic products of lipids as markers for the detection of irradiated meats, *Food Irradiation and the Chemist* (D. E. Johnston and M. H. Stevenson, eds.), Royal Society of Chemistry, Cambridge, United Kingdom, 1990, pp. 13-24.

154. M. H. Stevenson, Progress in the identification of irradiated foods. *Trends in Food Science and Technology* 3: 257-262 (1992).

155. B. J. Tabner and V. A. Tabner, Electron spin resonance spectra of gamma-irradiated citrus fruit skins, skin components and stalks. *Int. J. Food Sci. Technol.* 29: 143-152 (1994).

156. M. F. Desrosiers, F. G. Le, and W. L. McLaughlin, Inter-laboratory trials of the EPR method for the detection of irradiated meats containing bone *Int. J. Food Sci. Technol.* 29: 153-159 (1994).

157. S. Onori and M. Pantaloni, Electron spin resonance technique identification of irradiated chicken eggs. *Int. J. Food Sci. Technol.* 29: 671-677 (1995).

158. G. A. Schreiber, N. Helle, and K. W. Bögl, An interlaboratory trial on the identification of irradiated spices, herbs, and spice-herb mixtures by thermoluminescence analysis. *J. AOAC Int.* 78(1): 88-93 (1995).

159. B. Bhushan, R. M. Kadam, P. Thomas, and B. B. Singh, Evaluation on electron spin resonance technique for the detection of irradiated mango (Mangifera indica L.) fruits, *Int. J. Food Sci. Technol.* 29: 679-686 (1995).

160. N. Deighton, S. M. Glidewell, B. A. Goodman, and I. M. Morrison, Electron paramagnetic resonance of gamma-irradiated cellulose and lignocellulosic material. *Int. J. Food Sci. Technol.* 28: 45-55 (1993).

161. G. Wirtanen, S. Salo, M. Karwoski, and A. M. Sjöberg, A microbiological screening method for the indication of irradiation of frozen poultry meat. *Z. Lebensm. Unters Forsch* 200: 194-197 (1995).

162. D. J. Deeble, A. W. Jabir, B. J. Parsons, C. J. Smith, and P. Wheatley, Changes in DNA as a possible means of detecting irradiated food. *Food Irradiation and the Chemist* (D. E. Johnston and M. H. Stevenson, eds.), Royal Society of Chemistry, Cambridge, United Kingdom, 1990, pp. 57-79.

163. T. F. Rowe and K. J. Towner, Effect of irradiation on the detection of bacterial DNA in contaminated food samples by DNA hybridization. *Letters in Applied Microbiology* 18: 171-173 (1994).

164. P. H. G. Sharpe, Dosimetry for food irradiation, *Food Irradiation and the Chemist* (D. E. Johnston and M. H. Stevenson, eds.), Royal Society of Chemistry, Cambridge, United Kingdom, 1990, pp. 109-123.

165. E. A. Murano, ed., *Food Irradiation: A Sourcebook*, Iowa State University Press, Ames, Iowa, 1995.

166. G. H. Pauli and L. M. Tarantino, FDA regulatory aspects of food irradiation. *J. Food Prot.* 58(2): 209-212 (1995).

167. K. A. Gilles, R. E. Engel, and D. Derr, Impact of technological advances in food processing and preservation, in particular irradiation, on international food trade, Proceedings of the International Conference on the Acceptance, Control of and Trade in Irradiated Food, Geneva, 1988, pp. 73-86.

168. G. Moy, Foodborne disease and the preventive role of food irradiation. *IAEA Bulletin* 4: 39-43 (1992).

169. J. F. Diehl, *Safety of Irradiated Foods*, 2nd ed., Marcel Dekker, New York, 1995.

8

Chemicals and Biochemicals Used in Food Preservation

I. INTRODUCTION

The use of chemicals in foods is a well-known method of food preservation. Several excellent reviews, articles, and books are available in the literature describing the use of chemicals as food preservatives. This chapter is a brief overview of the use of chemicals and biochemicals as a nonthermal method of food preservation. Considering the vastness of the subject, we do not intend to describe it here in detail.

Additives are used in foods to perform one of the following functions: to preserve, add flavor, add color, improve the texture and/or the nutritional value of the food. Chemical preservatives are defined by the Food and Drug Administration as "any chemical that, when added to food, tends to prevent or retard deterioration, but does not include common salt, sugars, vinegars, spices or oils extracted from spices, substances added to food by direct exposure thereof to wood smoke or chemicals applied for their respective insecticidal or herbicidal properties" (1).

Foods are susceptible to chemical and biological deterioration. Preservatives are used to prevent deterioration. Chemical deterioration includes browning, lipid oxidation, and staling among others. Antibrowning agents, antioxidants, or antistaling agents are used to prevent chemical deterioration of the food. The biological deterioration of foods can be prevented using antimicrobials. By preventing or retarding chemical and biological deterioration, the shelf life of foods can be extended by as much as 200% (1).

The selection of a specific antimicrobial for a specific food is not an easy task, and many factors must be considered: physical and chemical properties of the antimicrobial agents, composition of the food, type of preservation system other than chemicals used in the food, characteristics and number of microorganisms, cost and toxicity of the antimicrobial.

The physical and chemical properties considered in selecting an antimicrobial are water solubility, hydrophile–lipophile balance, boiling point, ability to ionize, and potential interaction with the food constituents. The activity of antimicrobials is reduced as the result of reaction with lipids, proteins or carbohydrates. Further, chemical reactions between antimicrobials and food constituents may result in the formation of off-flavors and off-colors (1). Therefore, the antimicrobial used in a food should be carefully selected after considering the necessary factors.

Antimicrobials can be classified in the following categories: naturally present or formed in the food; chemicals added to the food; bacteriocins; and chemicals with multifunctional properties added to the food, one property being antimicrobial.

II. ANTIMICROBIALS NATURALLY PRESENT OR FORMED IN FOOD

A. Lactoperoxidase System (LPS)

Lactoperoxidase is the most abundant peroxidase in bovine milk. Milk also contains several antibodies and non-antibody proteins such as lysozyme, lactoferrin, and xanthine oxidase. Lactoperoxidase is a glycoprotein containing one heme group and 9.9–10.2% carbohydrate (2). The enzyme lactoperoxidase is present in milk in amounts varying between 10 and 30 µg/mL. For the LPS to function efficiently, 0.5–1.0 mg/mL of the enzyme lactoperoxidase is necessary (3). The peroxidases have no antimicrobial properties but can bring about the formation of potent antimicrobial compounds when combined with appropriate cofactors.

The enzyme lactoperoxidase forms the lactoperoxidase system (LPS) in combination with an oxidizable substrate and hydrogen peroxide (2). The oxidizable substrate may be a halide, thiocyanate -SCN⁻ (pseudohalide), or idican (indoxyl sulfuric acid). The anion SCN⁻ is found in stomach and kidney; in the mammary, salivary, and thyroid glands; and in the synovial, cerebral, spinal, lymph, and plasma fluids. Detoxification reactions between thiosulfates and metabolic products of sulfur amino acids or cyanide in the body result in the formation of SCN⁻. Hydrolysis of foods containing glucosides yields SCN⁻. Milk generally contains up to 15 µg/mL of thiocyanate (3).

Hydrogen peroxide is present at small concentrations of 2–4 µg/mL in fresh milk. H_2O_2 may also be generated by catalase-negative microorganisms under aerobic conditions or by the action of xanthine oxidase, glucose oxidase, ascorbic acid, and sulfhydryl oxidase in the presence of free oxygen (2).

The lactoperoxidase-catalyzed oxidation of thiocyanate is represented by the following reactions (2):

$$2SCN^- + H_2O_2 + 2H^+ \rightarrow (SCN)_2 + 2H_2O$$
$$(SCN)_2 + H_2O \rightarrow HOSCN + SCN^- + H^+$$
$$HOSCN \rightarrow OSCN^- + H^+$$

The thiocyanate anion may be oxidized to hypothiocyanate ($OSCN^-$) in a single step as (2):

$$SCN^- + H_2O_2 \rightarrow OSCN^- + H_2O$$

The hypothiocyanate oxidizes free sulfhydryl groups of bacterial cells to the corresponding disulfides, sulphenylthiocyanates, or sulphenic acids. These oxidation reactions disrupt the cell functions, causing damage to the cytoplasmic membrane and sugar and amino acid transport systems. However, a threshold concentration of $OSCN^-$ is required before significant inhibition occurs. The naturally occurring LPS is primarily active against H_2O_2-producing bacteria such as lactococci and lactobacilli. Microorganisms inhibited by the LPS include viruses, gram-positive and gram-negative bacteria, fungi, and possibly mycoplasmas and parasites. Gram-negative bacteria include coliforms, pseudomonads, salmonellae, and shigellae. Gram-positive bacteria, including streptococci and lactobacilli, are only temporarily inhibited. The gram-negative bacteria are more susceptible than the gram-positive (3).

The LPS can be used for preservation of milk, infant formula, and liquid whole eggs. The shelf life of pasteurized milk can be extended by 20 days with the use of LPS. When used in milk, the LPS is effective at temperatures below 15°C. A combination of the LPS and mild heat treatment may be an effective preservation method, particularly for minimally processed or heat-sensitive foods (3).

One of the drawbacks of the LPS is the affinity of lactoperoxidase enzyme to adhere to surfaces such as glass. Also, the LPS can be inactivated by reducing compounds and enzymes such as horseradish peroxidase. Enzymes competing for H_2O_2 can reduce the inhibitory activity of the LPS (3). The antibacterial effect of LPS is inhibited by sulfhydryl-containing compounds such as cysteine and by skim milk that received severe heat treatment (2).

B. Lysozyme

Lysozyme is a lytic enzyme naturally present in plant tissues, milk, and eggs. It is present in a concentration of 0.1 and 0.13 μg/mL in human and bovine milk, respectively (3). The bactericidal and lytic action of lysozyme depends on the enzyme concentration, ionic strength, and nature of the electrolytes present. Anions such as SCN^- and HCO_3^- promote bacterial lysis (2). Lysozyme destroys the cellular integrity via hydrolysis of the β (1–4) linkages between N-acetylmuramic acid and N-acetylglucosamine of the peptidoglycan layer of bacteria cell walls (3). Loss of the cell wall allows water uptake leading to cell lysis. Therefore, lysozyme is effective against gram-positive bacteria (2).

Gram-negative bacteria are more resistant to lysozyme because of the smaller content of peptidoglycan. Also the peptidoglycan is situated below the outer membrane (2). The outer membrane, consisting of a lipoprotein–lipopolysaccharide layer makes gram-negative bacteria less susceptible to damage by lysozyme. However, gram-negative bacteria that are sensitive to lysozyme include *Salmonella, Pseudomonas, Aeromonas,* enteropathogenic *E. coli,* and some *Vibrio* spp (3). Few of the gram-positive bacteria, including staphylococci, containing teichoic acids in the cell wall are resistant to lysozyme (2).

Although lysozyme may potentially be used in food systems for its preservative properties, it is used to a limited extent in the food industry. Lysozyme in combination with EDTA (ethylenediaminetetraacetic acid) or TRIS (tris [hydroxymethyl] aminomethane) can be used as a preservative in foods prepared from fresh vegetables (3). EDTA and TRIS act by binding divalent cations Ca^{2+} and Mg^{2+}, responsible for maintaining the structural integrity of the outer membrane of gram-negative bacteria (2). Lysozyme is approved as a food additive in Europe, Asia, and the United States. In Europe, lysozyme is added in certain hard cheeses to prevent gassing and blowing due to *Clostridium.* The Japanese use lysozyme to preserve oysters, shrimp, as well as other seafood, sushi, sake, noodles, potato salad, and custard (3).

C. Others

Lactoferrin is a glycoprotein present in milk that acts as an antimicrobial agent by binding the iron needed for growth of the microorganisms. Human milk contains up to 0.5 g of lactoferrin per 100 mL, whereas mature bovine milk contains very little lactoferrin. Lactoferrin (also known as lactotransferrin or lactosiderophilin) has two iron-binding sites. For each ferric ion bound by lactoferrin, one bicarbonate anion is bound and three protons are released

(3). The bacteriostatic effects of lactoferrin are expressed against bacteria with high iron requirements such as coliforms. Organisms with low iron requirements, such as lactic acid bacteria and streptococci, remain unaffected (2). Some of the bacteria inhibited by lactoferrin include *Listeria monocytogenes, Bacillus subtilis, Bacillus stearothermophilus, Staphylococcus aureus, Klebsiella,* and *E. coli* (2,3). The ability of lactoferrin to bind iron is inversely related to the citrate concentration. Citrate sequesters iron chelated to lactoferrin, and bacteria can consume the iron citrate complex. Human milk contains 2–3 mM of citrate whereas bovine milk contains 4–8 mM. Such citrate concentrations can inhibit the bacteriostatic activity of lactoferrin unless the bicarbonate concentration is also high (2).

Ovotransferrin, an albumen constituent of egg white, functions by sequestering available ferric ion. Ovotransferrin is one of the major components of the antimicrobial defense system in eggs. The ovotransferrin molecule possesses two iron binding sites; for each ferric ion bound, an anion is also bound. The anion may be carbonate, bicarbonate, oxalate, or EDTA. Ovotransferrin is more efficient in an alkaline environment of pH 9.5. Avidin, a glycoprotein present in egg white, deprives microorganisms of the essential vitamin biotin. It consists of four identical polypeptide subunits and is able to bind four biotin molecules, one per subunit (2). Conalbumin is another iron-binding protein present in egg white. Conalbumin constitutes about 10–12% of the total egg white solids. The binding of iron by conalbumin retards the rate of growth of microorganisms. Gram-positive bacteria are more sensitive to iron binding by conalbumin than are gram-negative bacteria. Of the gram-positive bacteria, micrococci and *Bacillus* are the most sensitive (3).

III. CHEMICALS WITH ANTIMICROBIAL PROPERTIES

A. Organic Acids

Acids are added to foods as acidulants or antimicrobials. The saturated acids with C10 to C12 chain lengths have optimal antimicrobial activity. The cis-forms of monounsaturated C16:1 and diunsaturated C18:2 have the highest activity. Gram-negative bacteria are affected by short ($< C6$), rather than long fatty acids and their derivatives. Organic acids inhibit the growth of microorganisms by inactivating or by affecting one or more of the following targets: cell wall, cell membrane, metabolic enzymes, protein synthesis system or genetic material. Organic acids are more effective antimicrobials than are inorganic acids (4).

When an undissociated acid molecule enters a living cell, it will dissociate because the internal pH is usually much higher than the pK of the acid. To maintain the pH of the cell, a compensating net transport of protons out of the cell has to take place. Alternatively, the acid molecule itself may be expelled from the cell by active transport. A constant extrusion of protons from the cell may cause depletion of cellular energy. Interference with energy metabolism was proposed as an explanation of the antibacterial action of short chain fatty acids (4).

Salts of benzoic, propionic, and sorbic acids are often used as antimicrobials. Other organic acids are primarily acidulants. One of the oldest chemicals used in the cosmetic, food, and drug industries is benzoic acid. The sodium salt of benzoic acid was the first chemical preservative approved by the U. S. Food and Drug Administration. Benzoic acid is also naturally present in cranberries, prunes, cinnamon, apples, strawberries, and yogurt. The antimicrobial properties of benzoic acid have been assumed to be expressed by the undissociated acid only. However, benzoic acid also shows a certain growth-inhibitory capacity in the dissociated state (4,5). At pH 3, 93.5% of the acid exists in the undissociated form; at pH 4, only 59.3% of the acid exists in the undissociated form (6). Benzoic acid and sodium benzoate are generally recognized as safe (GRAS) ingredients and are most suitable for foods with a pH < 4.5 (5).

Some of the foods preserved with benzoic acid and sodium benzoate include fruit juices, cider, carbonated or noncarbonated drinks, bakery products, salad dressings, margarine, tomato catsup, fruit salads, sauerkraut, jams, and jellies. Benzoic acid and sodium benzoate are also used as a fungicide to prevent postharvest diseases of fruits and vegetables. A benzoic acid–based polymer coating was developed to prevent spoilage of apples (5). Benzoate acts essentially as a mold and yeast inhibitor in acidic foods. The poor activity of benzoate at pH > 4.0 limits its use against bacteria. A undesirable "peppery" or bitter taste is imparted by benzoate when added to fruit juices at a concentration greater than 0.1% (7).

Benzoic acid blocks the oxidation of glucose and pyruvate at the acetate level and increases the oxygen consumption in the glucose oxidation in *Proteus vulgaris*. Benzoic acid inhibits the uptake of select substrate molecules such as amino acids (7). Benzoate influences enzymes controlling acetic acid metabolism and oxidative phosphorylation and also intervenes at various points in the tricarboxylic acid cycle, especially where the dehydrogenases of α-ketoglutaric acid and succinic acid are involved (4).

One of the causes of microbial growth inhibition due to benzoate is the elimination of the electrochemical gradient across the cell membrane. It was also suggested by some researchers (8) that benzoic acid influences the membrane either by interfering with the membrane proteins or changing the membrane fluidity. Benzoic acid also prevents the outgrowth of vegetative cell from spore after shredding of the spore wall (7).

Sorbic acid is a naturally occurring compound and is used in foods as the calcium, sodium, or potassium salt. The maximum permissible concentration of sorbates in foods is 0.2%. Sorbic acid is most effective at pH < 6.0. Sorbates are primarily effective against yeasts and molds. Bacteria inhibited by sorbates include *S. aureus*, *Salmonella*, coliforms, psychrotrophic spoilage bacteria, and *Vibrio parahaemolyticus*. Similar to benzoates, sorbates inhibit the amino acid uptake of cells and outgrowth of vegetative cells from spores. Sorbate enhances the heat activation and destruction of spores (8).

Enzymes such as malate, isocitrate, α-ketoglutarate, and succinate dehydrogenases, fumarase, and aspartase are inhibited by sorbate. It was suggested that sulfhydryl-containing enzymes, such as fumarase in catalase-positive bacteria, yeasts, and molds, aspartase, succinic dehydrogenase, and yeast alcohol dehydrogenase, react with sorbate. Sorbate at a concentration of 0.015–0.05% affects the cell wall, causing elongation with no division of cells after germination of *Bacillus* spores. Sorbate at a concentration of 0.4% inhibits synthesis of protein, RNA, and DNA in *Pseudomonas fluorescens* (4). Sorbates are mainly used in (3):

Dairy products: cheese, sour cream, yogurt
Bakery products: cakes, cake mixes, pies, fillings, toppings, doughnuts
Vegetables: fermented, pickles, fresh salads, relishes
Fruit: dried, jams, jellies, juices, salads, purees, concentrates
Beverages: still wines, carbonated or noncarbonated fruit drinks, low calorie drinks
Emulsions: mayonnaise, margarine, salad dressing
Meat and fish products: smoked and salted fish, dry sausages

In addition, sorbates are used as antibotulinal agents in meat products. Addition of nitrite to meat products is a health concern because of the formation of nitrosamines. The concentrations of nitrite added to meat products can be reduced by using a nitrite–sorbate combination. Sorbate is less toxic and four times as effective as benzoate or propionate (4).

Propionic acid has a slightly pungent and disagreeable odor. The salts of propionic acid exhibit a cheese-like flavor. The propionates have an effective upper pH limit of 5.5. Propionic acid is normally present in cheese and as a metabolite in the ruminant gastrointestinal tract. Propionic acid formed in Swiss cheese as a result of the growth and metabolism of propionic bacteria helps prevent mold growth. Propionic acid is used as a mold inhibitor in cheese foods and spreads, and as a mold and rope inhibitor in bread. The sodium salt of propionic acid is used in chemically leavened products because calcium interferes with the leavening action (9).

Acids, such as acetic, citric, propionic, malic, fumaric, and tartaric, are mainly used as acidulants, but they also inhibit molds. Organic acids can also act as chelating agents and thereby inhibit lipid oxidation, control pectin gel formation, inhibit browning, and aid in sucrose inversion (9).

Glacial acetic acid is a GRAS additive and is used as an acidifer and flavoring agent in sliced and canned fruits and vegetables, mayonnaise, salad dressings and sausages. Vinegar containing 5% acetic acid is more widely used in foods such as salad dressings, mustard, ketchup, marinades for meat, poultry and fish, and bakery products (10). Acetic acid is also effective in sanitization of meat carcasses and decreasing *Salmonella typhimurium* and *Campylobacter jejuni* in poultry scald water (9). Acetic acid is a better inhibitor of bacteria than of yeasts and molds. However, it is effective against bread molds *Aspergillus* and *Rhizopus* (4).

Citric acid is the most widely used organic acid in the food industry. It is used as a standard to evaluate the effects of other acidulants in foods. Citric acid is highly soluble in water and delivers a "burst" of tartness. In addition, citric acid can also chelate metal ions, allowing antioxidants to function more effectively in retarding lipid oxidation (10). Citric acid is particularly inhibitory to flat sour organisms from tomato juice. Other foods preserved with citric acid include ice cream, sherbets, beverages, salad dressings, fruit preserves, jams, and jellies. Citric acid is a precursor of diacetyl and therefore indirectly improves the flavor of cultured dairy products (9).

Adipic acid is the least acidic of the commonly used acidulants. Adipic acid functions as a buffering agent at pH 2.5–3.0, neutralizing agent, gelling aid, and sequesterant. Adipic acid is used in evaporated milk, instant puddings, gelatin desserts, cheese analogues, powdered concentrates for fruit-flavored beverages, cake mixes, baking powders, and flavoring extracts. The shelf life of dry mixes and baking powders is extended because of the extremely low moisture-absorption capacity of adipic acid (10).

Fumaric acid provides a strong acid taste and blends with certain flavoring compounds to intensify the aftertaste of select flavors. Fumaric acid is naturally present in rice, sugar cane, wine, plant leaves, mushrooms, and gelatin. The calcium, magnesium, potassium, and sodium salts of fumaric acid are approved for use in nonstandardized foods. However, FDA clearance is required for use of fumaric acid in foods covered by Standards of Identity. Similar to adipic acid, fumaric acid is used in dry mixes because of its strong resistance to moisture absorption. One advantage of fumaric acid is the smaller quantity of acid required to achieve a taste similar to citric, malic, or tartaric acids. Fumaric acid is used in fruit juice drinks, gelatin desserts, pie fillings, rye bread, and refrigerated biscuit doughs. Fumaric acid is used as an acidulant and clarifying agent in wine (10).

Tartaric acid has a strong, tart taste and enhances grape-like flavors. Tartaric acid is used in fruit jams, jellies, preserves, sherbets, and grape-flavored beverages. The monopotassium salt, commonly known as cream of tartar, acts synergistically with antioxidants to prevent rancidity and discoloration in cheese (9).

Lactic acid is a GRAS additive used in foods for acidification, flavor enhancement, and microbial inhibition. It is used in sugar confections, dairy products, meat products, beer and wine, beverages, bakery products and for calcium and sodium lactylates, which function as dough conditioners in baked goods (10).

Malic acid is a GRAS ingredient used in many foods, including nonalcoholic beverages, hard candies, canned tomatoes, and fruit pie fillings. It provides a smooth, tart taste that lingers in the mouth without imparting a burst of flavor. Due to its stronger apparent acidity, malic acid can be used in smaller amounts than citric acid (10).

The only inorganic acid widely used in the food industry is phosphoric acid. It provides the lowest pH of all food acidulants. It is used in foods for the purpose of acidification, as a buffer, to impart acidic taste, and to allow complexation of metal ions that promote product degradation. Phosphoric acid and a number of its salts have the GRAS status. Phosphoric acid is mainly used in carbonated beverages including root beer and cola. It is used for pH adjustment in cheese production and brewing, for controlling pH, and for binding metal ions in processing of vegetable fats and oils. It is used in jams and jellies as a buffering agent, as a chelator of metal ions that may dull the color of jelly, and to ensure strongest gel strength (10).

B.   Sulfur Dioxide and Sulfites

Sulfur may be added to foods in the form of sulfur dioxide, sodium and potassium salts of sulfite, bisulfite, or metabisulfite. With the exception of potassium sulfite, these salts are permitted for use in foods and beverages in the United States. In other countries, sulfites are permitted in liquid and dried egg products, fish fillets, fish paste, and sausage (6). The use of sulfites to preserve meat was known even in the early nineteenth century. The growth-inhibiting or lethal effects of sulfurous acid are most intense when the acid is in the undissociated form. The predominant ionic species of sulfurous acid depends on the pH, with SO_2 favored by pH < 3.0, HSO_3^- by pH between 3.0 and 5.0, and SO_3^{2-} by pH > 6.0 (7,11).

Bacteria are more sensitive to SO_2 than are yeasts and molds. SO_2 is bacteriostatic at small concentrations and bactericidal at large concentrations. Yeasts are intermediate to acetic acid and lactic acid bacteria and molds in their sensitivity to SO_2. Some strains of *Saccharomyces* produce SO_2 during juice fermentation (7). Proteins in the cell wall and plasma membrane of microorganisms are the primary targets interacting with sulfite. Sulfite ions react with a range of coenzymes essential for enzyme activity or with prosthetic groups conjugated with biologically active proteins (12).

Sulfite is added to foods and beverages for one or more of the following reasons: inhibition of enzyme activity, as an antioxidant, inhibition of nonenzymatic browning, inhibition of microbial growth, bleaching a foodstuff, and as a general aid in food processing. The bound forms of sulfur have little antimicrobial activity. Carbonyl compounds present in foods bind the sulfite, thus reducing the antimicrobial activity (12).

The major applications of sulfites in the wine making industry include sanitization of the equipment, inhibiting the growth of microorganisms from freshly squeezed juice, clarifying the wine during fermentation, and preventing microbial growth after fermentation (6). However, excess SO_2 can deplete thiamine and inhibit fermentation (11).

Sulfite treatment improves the color and odor of meats, and retards or prevents growth of surface bacteria. Sulfite treatment, combined with vacuum packaging and a good oxygen barrier film, extends the shelf life of sausage. SO_2 is the only legal fungicide for grapes. Gassing the fruit storage area with a high concentration of SO_2 completely inhibits mold growth accompanied by severe fruit discoloration. The use of small gassing concentrations three times a week gives better results than gassing once a week with high concentrations. SO_2 can also be used to control the growth of yeasts and molds on fruits such as apples, berries, and mulberries. Browning in fruits and vegetables can be controlled with the use of sulfites (11). Sulfites are also used to control microbial growth and maintain oxidative stability of carbonated beverages, dehydrated fruits, fruit juices, syrups, concentrates, purees, and fresh shrimp. Fruits and vegetables are sprayed with or dipped in a sulfite solution. Prior to canning sulfite-containing foods, sulfites must be removed to prevent generation of hydrogen sulfide and the black precipitate (6).

Thiamine pyrophosphate, a cofactor required for many enzymatic reactions, is readily destroyed by bisulfite. Because of the interaction with thiamine, sulfites are not allowed by law to be used in foods considered to be good sources of vitamin B_1 (6). Thiamine is cleaved by sulfite to produce pyrimidine sulfonate and a thiazole residue–containing product (12). The use of sulfur dioxide was generally recognized as safe (GRAS) until recently, when investigations indicated asthmatic individuals were placed at risk by relatively small amounts of sulfites (11). The alternatives to SO_2 include polyphenol oxidase inhibitors, ascorbic acid/citric acid–based formulations, and complexing agents such as EDTA. So far, no alternative with the multifunctional characteristics of SO_2 has been reported. In 1986, the FDA revoked the GRAS status of SO_2 for use on raw fruits and vegetables and established a residual concentration in select dehydrated foods. The presence of sulfiting agents must be declared on the label of standardized foods (13).

C. Nitrites and Nitrates

Until the early 1940s, nitrite was added to meats primarily to impart the characteristic flavor and color. The antimicrobial properties of nitrite were demonstrated by the mid-1950s. Nitrous acid derived from nitrite in the acid environment of meat is responsible for microbial inhibition. Nitrite is more effective as an antimicrobial agent at pH < 7.0. Nitrite was believed to prevent outgrowth of spores surviving after heat treatment in canned meat. Until the 1960s, it was believed that the antimicrobial effect of nitrite was caused by the influence of nitrite on water activity of the food. Although nitrite was recognized as an effective antimicrobial agent, the use of nitrite as a preservative in cured meats was not popular (14).

Curing was originally developed to aid in the preservation of food by the addition of sodium chloride. Sodium nitrite, a natural impurity of sodium chloride, was observed to be responsible for the development of pink-to-red pigment in cured meat. The adjuncts or additives used in curing may include ascorbates, nitrates, nitrites, phosphates, glucono-delta-lactone, and sugars. Fermentation, smoking, drying, and heating are aids to the curing process (15). The sodium salts of nitrite and nitrate are used in the curing formula. Sodium nitrite and nitrate stabilize the red meat color and inhibit certain spoilage and food poisoning microorganisms (7).

One of the disadvantages of using nitrite is poor stability to heat and storage. However, many bacteria are capable of utilizing nitrate as an electron acceptor, resulting in the reduction of nitrate to nitrite. In an acid environment, nitrite ionizes to yield nitrous oxide. The nitrous oxide further decomposes to nitric oxide. Nitric oxide reacts with myoglobin under reducing conditions to produce the red pigment nitrosomyoglobin (7).

The most important microorganism inhibited by nitrites is *Clostridium botulinum.* Nitrite is also effective against *C. sporogenes* and *C. perfringens*, the bacteria used in laboratory studies to assess potential antibotulinal effects of nitrites. The antibotulinal effect consists of inhibition of vegetative cell growth and prevention of germination and growth of spores viable after heat processing (7). Low concentrations of nitrite inhibit outgrowth of spores after germination whereas high concentrations inhibit the germination itself. For example, outgrowth of spores of *C. sporogenes* was inhibited with 1.4 mM of nitrite at pH 6.0, and germination was inhibited with 8.4 mM of nitrite at pH 6.0 (16). The antibotulinal activity of nitrite depends on the pH, salt content, temperature of incubation, and number of spores (7).

Nitrite makes iron of the heme groups unavailable for growth of *C. botulinum.* Nitric oxide rapidly and irreversibly inactivates the iron protein in the nitrogenase enzyme system of *C. pasteurianum* (14). Inhibition of the growth of *Clostridium* is enhanced in the presence of sequestering agents. The iron reacts with the sequestering agent and makes more nitrite available for conversion to nitric oxide (7).

An important source of ATP generation in clostridia is the oxidation of pyruvate to acetate by the phosphoroclastic system. Addition of nitrite to a *C. botulinum* cell suspension rapidly decreases intracellular ATP concentrations and promotes accumulation of pyruvate in the medium. The accumulation of pyruvate suggests inhibition of the phosphoroclastic system by nitrite. This system consists of two components, namely ferredoxin and pyruvate:ferredoxin oxidoreductase. While ferredoxin was inactivated to some extent, pyruvate:ferredoxin oxidoreductase is more sensitive to nitrite (16). Nitrite also inhibits oxygen uptake, glucose catabolism, and proline transport in *E. coli.* (14).

Because of the antibotulinal effect of nitrite, it is often necessary to add nitrite to cured meats. For the development of flavor and appearance, smaller concentrations (~100 ppm) of nitrite are required. However, the antibotulinal effect requires larger concentrations (~120 ppm) for bacon, comminuted cured ham, and canned, shelf-stable luncheon meat. Toxin production in meats can be greatly reduced by using a combination of nitrite, sucrose, and an inoculum of lactic acid bacteria (7). Only the required quantities of nitrite should be added to the food. The presence of excess concentrations of nitrite results in a green discoloration or nitrite burn in cured meats. In sausage, nitrite burn is caused by the conversion of nitrate to nitrite by micrococci near the periphery, and by *Staphylococcus* within the sausage (14).

Nitrite is also effective against *S. aureus* at large concentrations, with the effectiveness increasing as pH decreases. However, nitrite is generally ineffective against Enterobacteriaceae, including *Salmonella* and lactic acid bacteria. The antibacterial effects of nitrite may be the result of the reaction of nitric oxide with other porphyrin-containing compounds such as catalase, peroxidases, and cytochromes (7).

Two interesting observations were made in the 1960s: (a) Botulism was almost absent in cured, canned, and vacuum packaged meats and fish products; the meats exhibited no toxicity even in the presence of viable endospores and (b) about ten times greater concentration of nitrite was required to inhibit clostridia if the nitrite was added to a laboratory culture medium after autoclaving. These observations led to the conclusion that heating the medium with nitrite results in the formation of a substance about ten times more inhibitory than the nitrite. This substance is referred to as the Perigo factor. However, formation of the Perigo factor in cured meats is questionable. Also, the Perigo factor does not develop in all media. The Perigo factor requires heating to at least 100°C. Addition of meat to a medium containing the Perigo factor results in a loss of the inhibitory activity (7).

Use of nitrites in meats is controversial in terms of health risks because of the formation of carcinogenic nitrosamines. The reaction between nitrite and secondary, tertiary, or quaternary amines results in the formation of nitrosamines (7). A report by the National Academy of Sciences concluded that about 87% of ingested nitrate comes from vegetables. The ingested nitrate after conversion to nitrite represents 72% of the total intake, with less than 10% coming from cured meats (16). To reduce the risks of consuming nitrosamines, the USDA reduced the input NO_2 concentration for bacon to 120 ppm and set a maximum nitrosamine residual concentration of 10 ppb. The reduction in nitrite concentrations can be complemented with the use of sorbates, isoascorbate, or EDTA (7).

Nitrosothiol compounds were studied as possible replacements for nitrite. These compounds produce the desired color changes in meat and also act as antioxidants. However, the anticlostridial activity of nitrosothiol compounds is less than that of nitrite, and there is a likelihood of formation of nitrosamines. The disadvantage of using nitrite/sorbate combinations in meat is the allergic response in some consumers. Also, sorbate and nitrite may react to form nitroso compounds with mutagenic activity. Therefore, the nitrite/sorbate combination is not a suitable replacement for nitrite (16).

IV. CHEMICALS WITH MULTIFUNCTIONAL PROPERTIES

A. Spices and Essential Oils

Spices are common ingredients used for the primary purpose of adding flavors to foods. Besides adding flavor, some of the spices also inhibit the growth of microorganisms. Spices, being natural substances of plant origin, are more appealing to consumers than are synthetic food additives. Whole spices are more effective inhibitors than are spice extracts. In general, gram-positive organisms are more sensitive than gram-negative organisms. Cinnamon, clove, and mustard are considered strong inhibitors; black pepper, red pepper, and ginger are weak inhibitors for a variety of microorganisms. Allspice, bay leaf, caraway, coriander, cumin, oregano, rosemary, sage, and thyme represent medium inhibitors (17).

The antimicrobial activity of spices resides in the essential oil of the spice. Essential oils are defined as a group of odorous principles, soluble in alcohol and to a limited extent in water, consisting of a mixture of esters, aldehydes, ketones, and terpenes (2). The major components of the volatile essential oils of cinnamon, allspice, and cloves consist of eugenol (2-methoxy-4-allyl phenol) and cinnamic aldehyde. Cinnamon and allspice contain 0.5–1.0% and 4.5% volatile oils, respectively. Clove buds contain about 17% essential oil composed of 93–95% eugenol. The major volatile oil components of oregano, thyme, and savory are carvacrol, p-cymene, and thymol. Oregano contains up to 50% thymol; thyme contains 43% thymol and 36% p-cymene; and savory contains 30–45% carvacrol and 30% p-cymene. Other spices with antimicrobial activity include sweet marjoram, laurel, pimiento, coriander, anise, peppermint, caraway, cumin, fennel, celery, dill, mustard, rosemary, sage, and turmeric (3).

Onions contain two phenolic compounds, protocatechuic acid and catechol. Inhibition of microbial spoilage in mayonnaise-based delicatessen salads with onion extracts was demonstrated (2). Some of the other plant foods containing antimicrobial substances include garlic, horseradish, banana, sweet potato, cabbage, radish, and cereals (2). Allicin or diallylthiosulfinic acid is the antimicrobial substance present in garlic. Allicin in the concentrations of 1 in 85,000 in broth is bactericidal to a wide variety of gram-negative and gram-positive organisms. A 5% concentration of garlic extract in food inhibits *S. aureus*. A bactericidal effect is exhibited against *S. typhimurium* and *E. coli* with reconstituted dehydrated onion and garlic at concentrations of 1% and 5%, respectively. The crude juices and solvent extracts of onion and garlic also inhibit the growth of *B. subtilis, Pseudomonas pyocyaneus, Candida, Cryptococcus, Rhodotorula, Torulopsis, Trichosporon,* and aflatoxin-producing *A. flavus* and *A. parasiticus* (3).

Some food additives used as flavoring agents exhibit antimicrobial properties. The flavoring agents are generally more antifungal than antibacterial (18). Diacetyl imparts a buttery flavor and is more effective against gram-negative bacteria and fungi than against gram-positive bacteria. *1*-Carvone imparts a spearmint-like flavor and d-carvone imparts a caraway-like flavor; both are effective inhibitors of fungi at concentrations >1000 ppm. Phenylacetaldehyde imparts a hyacinth-like flavor and is inhibitory to *S. aureus* and *Candida albicans* at concentrations of 100 ppm and 500 ppm, respectively. Menthol imparts a peppermint-like flavor and is inhibitory to *S. aureus* at 32 ppm, and to *E. coli* and *C. albicans* at 500 ppm. Vanillin and ethyl vanillin inhibit fungi at concentrations > 1000 ppm (7). The sensitivity of molds to vanillin increases in the order of *Aspergillus niger, A. parasiticus, A. flavus*, and *A. ochraceus* when tested in fruit-based and potato dextrose agars at pH 3.5 and water activity (a_w) 0.98 (19).

Not only the essential oils, but the oleoresins of spices exhibit antimicrobial activity. For example, the oleoresin of cinnamon is inhibitory against yeasts including *Candida lipolytica, Debaromyces hansenii, Hansenula anomala, Kloeckera apiculata, Lodderomyces elongisporus, Rhodotorula rubra, Saccharomyces cerevisiae*, and *Torulopsis glabrata*. Allspice oleoresin impairs recovery of heated yeasts (3).

The mechanism of the antimicrobial activities of spices is not well established, because there are few standard procedures to test the antimicrobial activities. Antimicrobial action due to impairment of a variety of enzyme systems involved in energy production or structural component synthesis was suggested (2). Some spices exhibit antimicrobial activities under certain conditions, but not others (3). The antimicrobial activity of spices depends on the test medium, spice, and the microorganism. The test medium may be water, microbiological medium, or a food or beverage. The antimicrobial activity of a spice depends on whether the spice is added as is, as an extract, or applied to the agar plate. The concentration and origin of the spice are also important factors. Antimicrobial activity may differ with strains within the same genera of microorganisms. Spores and vegetative cells exhibit different sensitivities to the spices (17). Therefore, there is a need to specify the procedure used to test the antimicrobial activity of the particular spice.

One of the commonly used food additives is salt. Addition of salt disturbs the osmotic balance of the food environment, leading to loss of water from the microbial and food cell. A high concentration of salt exerts a drying effect on both food and microorganisms. Loss of water from microbial cells (plasmolysis) results in cell inactivation (7). The amount of salt needed to prevent microbial growth depends on the type and growth stage of the microorganisms, type and properties of the food, and storage temperature. Xerophilic molds, osmophilic yeasts and halophilic bacteria can grow at low a_w and require salt for their growth. Most food-borne pathogenic bacteria do not grow at a_w below 0.92 or at salt concentration greater than 10%. As the acidity of the food increases, amount of salt needed decreases. Toxin production by *C. botulinum* is delayed at low temperatures in bologna containing 2.2% salt. On the other hand, in the presence of increased salt concentration (> 0.09 M), growth of *S. aureus*, enterotoxin production, and coagulase and thermostable nuclease formation occurred at temperatures up to 2°C above those for growth in the presence of regular (0.09 M) amounts of salt (20).

Salt adds flavor and inhibits pathogenic bacteria in meat products. A small increase in salt concentration played a major role in delaying the formation of toxin from *C. botulinum* in meat products. For example, no toxin was produced in bologna with a salt concentration of 3.0% until four weeks when stored at 30°C, whereas toxin was produced in one week when the salt concentration was 2.2%. Vacuum-packaged meats and sausages containing less than 4% salt showed little protection from *C. botulinum* even in the presence of 50 or 100 ppm nitrite.

An appropriate concentration of salt in vegetable fermentations extracts nutrients from the plant tissue and allows the correct progression and growth of lactic acid bacteria. Acid produced by the lactic acid bacteria suppresses the growth of spoilage and pathogenic bacteria. Lower concentration of salt in fermented foods results in poor quality and even unsafe products. Salt exhibits synergistic antimicrobial effects with benzoate, sorbate, phosphate, antioxidants, spices, liquid smoke, and isoascorbate (20).

Sugar exerts antimicrobial action in a manner similar to salt and is used in many food preparations; the last chapter of this book presents several examples. The only difference between the use of salt and sugar is the high concentration of sugar (six times) needed for the same effect as salt (7).

B. Antioxidants

Antioxidants are defined by the Food and Drug Administration (FDA) as substances used to preserve food by retarding deterioration, rancidity, or discoloration due to oxidation. The inhibitory effect of antioxidants is attributed to the donation of electrons or hydrogen to a fat containing free radical and to the formation of a complex between the antioxidant and the fat chain. The primary function of antioxidants is to prevent auto-oxidation of lipids in foods. For effective protection of the fat in foods, antioxidants must be added as early as possible in the manufacturing process. Antioxidants can be added also to the finished product. Antioxidants cannot reverse the oxidation of rancid oils, nor are they effective in suppressing enzymatically induced oxidation (21).

Butylated hydroxyanisole (BHA) is more effective in suppressing oxidation of animal fats than vegetable fats. BHA is also effective in protecting the color and flavor of essential oils. The oxidation of coconut and palm kernel oils used in cereal and confectionery products is controlled by using BHA. Tertiary butylated hydroxyquinone (TBHQ) is the best antioxidant for protecting frying oils against oxidation and also provides carry-though protection to the finished product. A combination of TBHQ and a chelating agent such as citric acid enhances the antioxidant properties. Tocopherols are natural antioxidants present in plant tissues and are effective in bacon, baked goods, butter fat, lard, margarine, rapeseed oil, safflower oil, and sunflower seed oil. Tocopherols have GRAS status and are used as food preservatives (21).

Antioxidants such as BHA, butylated hydroxytoluene (BHT) and TBHQ exhibit the secondary function of inhibiting gram-positive and gram-negative bacteria, yeasts, and molds. Food-borne pathogens such as *B. cereus*, *V. parahaemolyticus*, *Salmonella*, and *S. aureus* are effectively inhibited at concentrations of < 500 ppm (7).

V. BACTERIOCINS

Bacteriocins may be defined as protein-containing macromolecules with a capacity to exert bactericidal action on susceptible bacteria. *E. coli* produce colicins and lactococci produce nisin and diplococcin. The bacteriocins inhibit some gram-positive bacteria, spore-forming bacteria, and food-borne pathogens, including *Listeria monocytogenes* (22).

Nisin is produced by fermentation of a modified milk medium by certain strains of lactic acid bacterium, *Lactococcus lactis*. The potential applications for nisin in food preservation were observed when nisin-producing starter cultures prevented clostridial gas formation in cheese. The nisin molecule is acidic in nature and exhibits the greatest stability under acidic conditions (23). Some of the desirable properties of nisin include nontoxicity and heat and storage stability; it is a naturally produced additive, is destroyed by digestive enzymes, has a narrow range of antimicrobial activity, and does not present formation of off-flavors or off-colors (7).

Nisin possesses antimicrobial properties against certain gram-positive bacteria, including *Staphylococcus*, *Streptococcus*, *Micrococcus*, and *Lactobacillus*. Gram-negative bacteria, yeasts, and molds are not affected by nisin. Spores of *Bacillus stearothermophilus* are more sensitive than spores of *B. cereus*, *B. polymyxa*, or *B. megaterium*. Spores of *C. botulinum* types A, B, and E are sensitive to nisin. The sensitivity of the spores decreases in the order of type E, B, and A. Lower pH, higher temperature, and a longer period of heat shock may be used to increase the effectiveness of nisin in preventing the outgrowth of C. *botulinum* spores (23).

Conventional heat treatment of low-acid canned food requires an F_o treatment of 6 to 8 min to inactivate spores of C. *botulinum* and spoilage organisms. The addition of nisin to low acid foods helps reduce the F_o to 3 min. The reduced time and temperature results in better quality food (7).

Nisin disrupts the cytoplasmic membrane, resulting in the leakage of ATP from the cell or, in more severe cases, cell lysis. Nisin inactivates the sulfhydryl groups in the cytoplasmic membrane (21). Nisin also inhibits murein synthesis in the cell wall (7).

Other antibiotics include natamycin, tetracycline, subtilin, and tylosin. Natamycin, also known as pimaricin, tennecetin, and myprozine, is effective against yeasts and molds but not bacteria. The concentration of natamycin required to inhibit certain strains of molds is 50–100 times less than the inhibitory concentration of sorbic acid. Compared to nystatin, about five times less natamycin is required to control fungi on strawberries and raspberries. Natamycin binds to membrane sterols and induces distortion of selective membrane permeability. The lack of membrane sterols in bacteria explains the ineffectiveness of natamycin. The surface treatment of refrigerated meats with 7 to 10 ppm of chlortetracycline (CTC) or oxytetracyline (OTC) typically results in a shelf life extension of 3 to 5 days. CTC combined with sorbates can prevent spoilage of fish up to 14 days. Heat sensitivity and lack of storage stability are the disadvantages of tetracycline. Subtilin, produced by some strains of B. *subtilis*, is similar to nisin in its action against gram-positive bacteria and prevention of outgrowth of germinating spores. One ppm of tylosin added to cream-style corn containing flat sour spores, followed by thermal process, resulted in the extension of shelf life (at 54°C) up to 30 days (7).

VI. SUMMARY

Antimicrobials may be naturally present or added to foods. The naturally present antimicrobials and organic acids exhibit good inhibitory properties. However, some of the chemicals used as preservatives are controversial in terms of the health risks involved and the benefits of extending the shelf life of the food. Attempts are being made to replace the controversial chemicals with new antimicrobials or combinations of safe chemicals.

A few food constituents or preservatives, such as spices and antioxidants, exhibit the multifunctional properties of adding flavor or specific taste to the food and acting as antimicrobial agents. One of the drawbacks of using a spice as a preservative is the lack of a standard procedure to evaluate the effectiveness of the spice as an antimicrobial agent. The antimicrobial effectiveness of the spice is dependent on select and circumstantial factors such as concentration, origin, and pH. A proper study on the use of spices as antimicrobial agents will be helpful in avoiding or reducing the risk associated with the addition of chemical preservatives to foods.

REFERENCES

1. A. L. Branen, Introduction to use of antimicrobials, *Antimicrobials in Foods* (P. M. Davidson and A. L. Branen, eds.), Marcel Dekker, New York, 1993, p. 1.
2. K. M. Wilkins and R. G. Board, Natural antimicrobial systems, *Mechanisms of Action of Food Preservation Procedures* (G. W. Gould, ed.), Elsevier, London, 1989, p. 285.
3. D. E. Conner, Naturally occurring compounds, *Antimicrobials in Foods* (P. M. Davidson and A. L. Branen, eds.), Marcel Dekker, New York, 1993, p. 441.
4. T. Ecklund, Organic acids and esters, *Mechanisms of Action of Food Preservation Procedures* (G. W. Gould, ed.), Elsevier, London, 1989, p. 161.
5. J. R. Chipley, Sodium benzoate and benzoic acid, *Antimicrobials in Foods* (P. M. Davidson and A. L. Branen, eds.), Marcel Dekker, New York, 1993, p. 11.
6. J. D. Dziezak, Preservatives: antimicrobial agents. *J. Food Technol.* 40 (9): 104 (1986).
7. J. M. Jay, Food preservation with chemicals, *Modern Food Microbiology*, Van Nostrand Reinhold, New York, 1992, p. 251.
8. J. N. Sofos and F. F. Busta, Sorbic acid and sorbates, *Antimicrobials in Foods* (P. M. Davidson and A. L. Branen, eds.), Marcel Dekker, New York, 1993, p. 49.
9. S. Doores, Organic acids, *Antimicrobials in Foods* (P. M. Davidson and A. L. Branen, eds.), Marcel Dekker, New York, 1993, p. 95.
10. J. D. Dziezak, Acidulants: Ingredients that do more than meet the acid test. *J. Food Technol.* 44(1): 76 (1990).
11. C. S. Ough, Sulfur dioxide and sulfites, *Antimicrobials in Foods* (P. M. Davidson and A. L. Branen, eds.), Marcel Dekker, New York, 1993, p. 137.
12. A. H. Rose and B. J. Pilkington, Sulfite, *Mechanisms of Action of Food Preservation Procedures* (G. W. Gould, ed.), Elsevier, London, 1989, p. 201.
13. J. Giese, Antimicrobials: assuring food safety. *J. Food Technol.* 48 (6): 102 (1994)
14. R. B. Tompkin, Nitrite, *Antimicrobials in Foods* (P. M. Davidson and A. L. Branen, eds.), Marcel Dekker, New York, 1993, p. 191.
15. M. K. Wagner and L. J. Moberg, Present and future use of traditional antimicrobials. *J. Food Technol.* 43 (1): 143 (1989).
16. L. F. J. Woods, J. M. Wood, and P. A. Gibbs, Nitrite, *Mechanisms of Action of Food Preservation Procedures* (G. W. Gould, ed.), Elsevier, London, 1989, p. 225.

17. L. L. Zaika, Spices and herbs: their antimicrobial activity and its determination. *J. Food Safety* 9: 97 (1988).

18. J. M. Jay and G. M. Rivers, Antimicrobial activity of some food flavoring compounds. *J. Food Safety* 6: 129 (1984).

19. A. López-Malo, S. M. Alzamora, and A. Argaiz, Effect of natural vanillin on germination time and radial growth rate of moulds in fruit based agar systems. *Food Microbiology* 12: 213 (1995).

20. J. N. Sofos, Antimicrobial effects of sodium and other ions: a review. *J. Food Safety* 6: 45 (1983).

21. J. D. Dziezak, Antioxidants. *Food Technol.* 40: 94 (1986).

22. M. A. Daeschel, Antimicrobial substances from lactic acid bacteria for use as food preservatives. *Food Technol.* 43 (1): 164 (1989).

23. J. D. Broughton, Nisin and its use as a food preservative. *Food Technol.* 44 (11): 100 (1990).

9

Hurdle Technology

I. INTRODUCTION

The major food preservation techniques are based on the delay or prevention of microbial growth, using factors that most influence the growth and survival of microorganisms, such as temperature, water activity (a_w), oxidation-reduction potential, pH, available substrates, presence or absence of oxygen, concentration of major solutes present, and preservatives. The use of the inhibiting factors in combination can be advantageous principally by allowing the less extreme use of any single treatment (1). Leistner (2,3) introduced the hurdle effect, to illustrate that in most traditional and novel foods, a combination of several preservation factors (hurdles) that should not be overcome by the microorganisms present account for the final microbial stability and safety of food. The hurdle concept illustrates that complex interactions of temperature, a_w, pH, and so on, are significant to the microbial stability of foods. The application of the synonymously called combined methods, combined processes, combination preservation, combination techniques, barrier technology, or *métodos combinados* in Spanish, *tecnologia degli ostacoli* in Italian, *Hürden-Technologie* in German, *technologie des barrières* in French, and *hurdle technology* in English potentially improves product quality and identifies innovative food product possibilities (4).

The spoilage and contamination of foods by microorganisms is an important problem worldwide despite the wide range of preservation techniques employed (5). Current consumer demand for foods that are less severely processed, additive-reduced, natural, and fresh prompts food manufacturers to select milder preservation techniques. The hurdle technology concept can address these needs. Up to now, about fifty different hurdles have been identified in food preservation (5). The commercial challenge of foods minimally processed provides a strong motivation to study food preservation systems combining traditional microbial stress factors or hurdles, while introducing "new" variables for microbial control such as high pressure, high intensity pulsed electric fields, oscillating magnetic fields, light pulses, food irradiation, chemicals and biochemicals. An improved understanding of the mechanisms underlying the effectiveness of the nonthermal processes and the combinations with the traditional hurdles is therefore urgently required, so that the new preservation possibilities can move forward with a sound scientific basis because, most likely, combining technologies is the future of food preservation.

II. NONTHERMAL METHODS AS HURDLES

Heat treatment is the most utilized method for stabilizing foods because of its ability to destroy microorganisms and inactivate enzymes. However, because heat can alter many organoleptic properties of foods and diminish the bioavailability of some nutrients, there is a growing interest in searching for methods able to reduce the intensity of heat treatments needed for food preservation (6).

The inactivation of microorganisms in high intensity electric fields is related to changes in the cell membrane due to an electromechanical instability. Vega-Mercado et al. (7) examined the possible synergistic effect of pulsed electric fields, pH, and ionic strength in the inactivation of *Escherichia coli* at temperatures ranging from 10° to 15°C. The electric field and ionic strength are more likely related to the poration rate and physical damage of the cell membranes; pH is related to changes in the cytoplasmic conditions due to the osmotic imbalance caused by poration. Liu et al. (8) studied the inactivation of *E. coli* O157:H7 by the combination of antimicrobial organic acids and pulsed electric fields. Inactivation was more pronounced for benzoic and sorbic acids than for acetic acid. Increases in field intensity and number of pulses increased the extent of inactivation. The combination of heat, lysozyme, and high intensity pulsed electric fields could be an alternative for the sterilization of food products as reported by Simpson et al. (9). In the context of this chapter, pulsed electric fields can be considered a hurdle that combined with additional factors—such as pH, ionic strength, temperature, and antimicrobial agents—can be used effectively in food preservation.

High pressure presents unique advantages over conventional thermal treatments (10). However, many reported data show that commercial pasteurization or sterilization of low-acid foods using high pressure is very difficult without some additional factors to enhance the inactivation rate. Factors such as heat, antimicrobials, ultrasound, and ionizing radiation can be used in combination with high pressure to accelerate the rate of inactivation.

High pressure can be used to reduce the severity of the factors traditionally used to preserve foods. The use of high pressure in combination with mild heating has considerable potential (11). Studies have shown that the antimicrobial effect of high pressure can be increased with heat, low pH, carbon dioxide, organic acids, and bacteriocins such as nisin (12–14). Therefore, the hurdle concept could be applied to the optimization of high hydrostatic pressure for low-acid foods; a combination of moderate treatments including pressure can lead to a food preservation method effective against bacterial spores (13).

Popper and Knorr (15), Papineau and Schmersal (16), and Knorr (10,17) reported enhanced pressure inactivation of microorganisms when pressure treatments were combined with additives such as acetic, benzoic, or sorbic acids, sulfites, some polyphenols, and chitosan; these combination treatments allow lower processing pressure, temperature, and/or time of exposure. Roberts and Hoover (13) evaluated the effect of combinations of pressure at 400 MPa, heat, time of exposure, acidity, and nisin concentration against *Bacillus coagulans* spores. Spores of *B. coogulans* sublethally injured by pressurization in combination with heat and acidity became more sensitive to nisin; the investigators concluded that acidic foods could be protected from spore outgrowth with the combined treatment. Hauben et al. (18) studied the lethal inactivation and sublethal injury of *E. coli* by high pressure and by combinations of high pressure treatments with lysozyme, nisin, and/or EDTA. High pressure treatments (180 to 320 MPa) disrupt the bacterial cells outer membrane, causing perisplamic leakage and sensitization to lysozyme, nisin, and EDTA; this demonstrates that sublethal injury can be usefully applied in a combined-methods approach as an effective food preservation method.

Crawford et al. (19) evaluated the combination of high hydrostatic pressure, heat, and irradiation to eliminate *Clostridium sporogenes* spores in chicken breast. No significant differences were reported in the number of surviving spores between samples that were first irradiated and then pressurized, or vice versa. However, significant differences were reported between samples exposed to combined treatments and those that were only irradiated, the combined process being more effective. Total inactivation of the initial inoculated spores (no survivors) was accomplished with a 6-kGy dose followed by pressurization at 690 MPa and 80°C for 20 min. A combination of lower doses of irradiation and high pressure is more useful in eliminating *C. sporogenes* spores than the application of either process alone.

Earnshaw et al. (20) mentioned that there is no synergistic antimicrobial action between sorbic acid and pressure up to 400 MPa applied to *Zygosaccharomyces bailii*, and attributed this lack of synergy to the modification of the sorbic acid dissociation constant under pressurization. Tauscher (21) reported that the carboxylic acids commonly used as food preservatives show enhanced ionization when subjected to high pressure. However, Palou et al. (22) demonstrated that increased antimicrobial effects can be obtained when combining high pressure and potassium sorbate to inactivate *Z. bailii* in laboratory model systems with reduced a_w and pH. The initial inoculum (10^5 *Z. bailii* CFU/mL) was completely inactivated in systems with a_w 0.98 in the presence of potassium sorbate with pressure ≥ 345 MPa for more than 2 min; without potassium sorbate, the pressure can be applied at 517 MPa for 4 min. In laboratory systems with a_w 0.95 and without potassium sorbate, the pressure must be ≥ 517 MPa for 10 min; and with potassium sorbate, the treatment time could be reduced to 4 min (22).

For most of the possible combined processes, the primary goal consists of identifying the factors or treatments that could sensitize microorganisms to pressure (23) or that could cause microbial death in sublethally pressure-injured microbial cells. However, the protective effects that could be exerted by food components made it necessary to assess each combination process for a particular food product.

Nonthermal processes show excellent promise for incorporation into combined preservation systems using the hurdle technology concept. The preservation systems have to be designed with an in-depth understanding of how the nonthermal processes work. Further research is needed to assure the commercial success of these nonthermal preservation systems.

III. MICROBIAL STABILITY OF FRUITS PRESERVED BY HURDLE TECHNOLOGY

Intermediate-moisture foods (IMF) have a moisture content higher than their dry counterparts and are stable without refrigeration. The a_w of IMF is generally reported to be in the range of 0.85–0.60. Chirife and Favetto (24) reported that the upper limit (0.85) corresponds to the minimal a_w for growth of *Staphylococcus aureus*. Solutes and humectants for adjusting the a_w of an IMF to the value of 0.85 are limited to some sugars, salts, and polyols. The quantities of the solutes needed to reach a_w 0.85 impart, in many cases, undesirable sensory characteristics and affect the physical properties of the food product. However, in many cases a mild reduction of a_w in combination with other mild processes or factors produces food products that are microbiologically stable (25). The hurdle technology concept in food preservation was first introduced for meat products (2). However, it has been succesfully applied to several other products, as reported by Leistner (4,26–28). The shelf-stable foods have an a_w of 0.90–0.95 and require additional hurdles or preservation factors to ensure microbial stability (26). Numerous preservation methods are identified as potential hurdles (26,29) and could be used for food preservation using the hurdle technology approach. Leistner and coworkers studied for many years shelf-stable and IMF products based on meat. Leistner (26) defined shelf-stable meat products (SSP) as high-moisture meats ($a_w > 0.90$) storable at room temperature after a mild heat treatment (70–110°C core temperature) in sealed containers. The process offers the following advantages: the mild heat treatment improves the sensory and nutritional properties of the SSP; the lack of refrigeration simplifies distribution and saves energy during storage; the required a_w is not as low for IMF, and thus fewer solutes or less drying of the product is necessary (26). Depending on the hurdle that exerts a determinant effect on product stability, Leistner (26,27) distinguishes F-SSP, a_w-SSP, and pH-SSP: the microbial stability of the meat product mainly relies on a mild thermal treatment (F-value), a_w reduction, or in the increased acidity (pH) of the product, respectively. However, additional hurdles and/or the combination of the hurdles (including a_w, pH, and heat) contribute to the sound shelf life of the meat products. Thus, the SSP meats in which stability is the result of a combination of factors—none of which exerts a major effect—were called Combi-SSP (4,26).

The principle used by Leistner et al. (30) for shelf-stable high-moisture meats ($a_w > 0.90$)—in which only a mild heat treatment is used and still the products exhibit long shelf life without refrigeration—can be applied to other foodstuffs. Because fruits serve as a good example, for the rest of this chapter, we discuss the preservation of fruits using hurdle technology to further understand the possibilities and/or opportunities of this approach.

In the case of IMF, the a_w restricts potential microbial spoilage to a narrow list of organisms, mainly fungi, that tolerate low a_w. Then, in order for IMF to be stable without refrigeration the addition of preservatives provides the safety margin. Many of the considerations on the significance of microorganisms in IMF are made in terms of a_w limits for growth. However, microbial control in IMF depends not only on a_w, but on pH, redox potential, thermal treatments, preservatives, and so forth, which also exert an important effect on the flora (31). Leistner and Rödel (2) indicated that the pH of IMF should be as low as palatability permitted and, whenever possible, below pH 5.0. However, the number of foods in which this can be applied is limited because pH cannot be reduced without flavor impairment.

Fruits are a good example of foods that can tolerate pH reduction without affecting flavor significantly. Important developments in IMF preservation based on fruits and vegetables are reported (32–34). The understanding of the principles that assure the microbial stability of many intermediate-moisture fruits—and the knowledge of the factors or hurdles and their required levels—can be used to develop new fruit products based on selected hurdles. In the case of fruits, if fresh-like quality is the goal, a_w reduction by addition of humectants should be employed at a minimum level to maintain the product in the high moisture range (35). To compensate, in terms of stability, for the high moisture of the product, a blanching treatment can be applied with minimal effects on sensory and nutritional properties; pH reductions that will not result in flavor impairment can be employed; and, finally, permitted preservatives can be used to reduce the risk of spoilage. The combination of the mentioned factors, placed in the context of the hurdle technology principles applied to fruits, make up an interesting alternative to traditional IMF fruits (31,35).

For the selection of the hurdles and their levels, it is necessary to know the characteristics of the fruit to be processed and the characteristics of the desired product. Based on the hurdle technology, a combination of factors that control microbial growth and maintain the fruit's organoleptic characteristics must be selected. The large majority of fruits are high-acid products, though certain fruits have a high pH (e.g., banana, melon, mamey, fig, and papaya). The low pH and the nature of the organic acid molecule select for the growth of acid-tolerant microorganisms, such as fungi (predominantly molds) and lactic acid bacteria (36). Based on this and on consumer demand for more fresh-like products, for the new technologies for fruit preservation developed as part of two multinational Iberoamerican projects the hurdle a_w was selected to be in the range 0.93–0.98 and the pH was maintained equal to or near the pH value of the fresh fruit (pH 3.0–4.1) (36). Table 1 presents some of the fruits preserved by these combined techniques applying the hurdle concept and the inhibition factors used. Figures 1 and 2 are examples of the high-moisture fruit products developed. In the fruits with higher pH, the latter was adjusted to the minimum value compatible with the natural flavor of the fruit. Incorporating preservatives, such as weak lipophilic acids (i.e., sorbic or benzoic acid) provides an additional hurdle for most spoilage fungi. The slight thermal treatment applied as a blanching step to inactivate enzymes inactivates or injures microorganisms, reducing the initial microbial load. Sulfiting

TABLE 1 Innovative Technologies for Fruit Preservation

Fruit	Inhibition factors	Storage temperature (°C)	Shelf life (months)	Reference
Slices or whole				
Peach, halves	Blanching (vapor, 2 min) $a_w = 0.98$ (sucrose) pH = 3.7 $NaHSO_3$ = 150 ppm KS = 1000 ppm	35	3	61
Peach, halves	Blanching (vapor, 2 min) $a_w = 0.94$ (glucose) pH = 3.5 $NaHSO_3$ = 150 ppm KS = 1000 ppm	20 or 30	4	46
Pineapple, sliced or whole	Blanching (vapor) $a_w = 0.97$ (glucose) pH = 3.1 $NaHSO_3$ = 150 ppm KS = 1000 ppm	27	4	49
Pineapple, sliced	Blanching (vapor) $a_w = 0.97$ (sucrose) pH = 3.8 $NaHSO_3$ = 150 ppm KS = 1000 ppm	35 25	3 8	36
Mango	Blanching (vapor, 4 min) $a_w = 0.97$ (sucrose) pH = 3.0 $NaHSO_3$ = 150 ppm KS = 1000 ppm	35	4.5	56
Papaya	Blanching (vapor, 30 sec) $a_w = 0.98$ (sucrose) pH = 3.5 $NaHSO_3$ = 150 ppm KS = 1000 ppm	25	5	48
Papaya	Blanching (vapor, 4 min) $a_w = 0.98$ (sucrose) pH = 3.5 $NaHSO_3$ = 150 ppm KS = 400 ppm	35	3.5	36

TABLE 1   Continued

Fruit	Inhibition factors	Storage temperature (°C)	Shelf life (months)	Reference
Fig	Blanching (2% w/w NaHCO$_3$, 80C, 5 min) a$_w$ = 0.95 (sucrose) pH = 3.0 KS = 1000 ppm	35	3	36
Strawberry	Blanching (vapor, 1 min) a$_w$ = 0.97 (sucrose) or 0.95 (glucose) pH = 3.1 AA = 200 ppm NaHSO$_3$ = 150 ppm KS = 1000 ppm	25	4	36
Pomalaca	Blanching (85°C, 5 min) a$_w$ = 0.97 (sucrose) pH = 3.5 SO$_2$ = 180 ppm KS = 1,300 ppm Hot filling	35	3	36
Purees				
Banana	Blanching (vapor, 1 min) a$_w$ = 0.97 (glucose) pH = 3.4 AA = 250 ppm NaHSO$_3$ = 400 ppm KS = 100 ppm Mild heat treatment (100°C, 1′)	27	3.5	50
Mango	Blanching (80°C, 10 min) a$_w$ ≈ 0.985[a] pH = 3.6 SMB = 150 ppm SB = 1000 ppm	30–35	3	36
Papaya	Blanching (vapor, 3 min) a$_w$ = 0.98 (sucrose) pH = 4.1 KS = 1000 ppm	35	4	36

TABLE 1 Continued

Fruit	Inhibition factors	Storage temperature (°C)	Shelf life (months)	Reference
Plum	Blanching (vapor, 3 min) $a_w = 0.98$ (sucrose) pH = 3.0 KS = 1000 ppm	25	4	36
Passion fruit	Blanching (vapor, 3 min) $a_w = 0.98$ (sucrose) pH = 3.0 $SO_2 = 150$ ppm KS = 400 ppm	30	4	36
Passion fruit	$a_w = 0.94$ (sucrose) pH = 3.4 Heat treatment (85°C, 2 min) $Na_2S_2O_3 = 150$ ppm KS = 1500 ppm Hot filling (60°C)	35	6	36
Tamarind	$a_w = 0.96$ (sucrose) pH = 2.5 Heat treatment (85°C, 2 min) $Na_2S_2O_3 = 150$ ppm KS = 1500 ppm Hot filling (60°C)	35	6	56

a_w = water activity; KS = potassium sorbate; AA = ascorbic acid; SB = sodium benzoate; SMB = sodium metabisulfite.
[a] a_w of the fresh fruit.
Source: From Ref. 36.

agents were used in low concentrations when necessary to inhibit nonenzymatic browning reactions (31,36,37).

Foods with high a_w are suitable habitats for the growth of bacteria, yeasts, and molds. In high-moisture food products (HMFP), the a_w values applied no longer represent a major constraint to growth of spoilage and pathogenic microorganisms; some HMFP can be stabilized at a_w values as high as 0.98. The pH will exert a strong selective pressure on the existing microflora (31,37). However, high-acid contents provide an unsuitable environment for the growth of most bacteria; therefore, the low pH of these minimally processed fruits establishes that the potential source of spoilage is yeasts,

FIGURE 1   Minimally processed high-moisture fruit slices (from left to right, apples, papayas, pineapples, mangoes, and peaches). [Courtesy of Universidad de las Américas-Puebla, Mexico.]

FIGURE 2   Minimally processed high-moisture banana purees in assorted packages. [Courtesy of Universidad de las Américas-Puebla, Mexico.]

molds, and acid-tolerant bacteria. Interactions pH–a_w in the applied ranges will be enough to suppress the growth of most bacteria of concern in fruit preservation (31,36). At high levels of a_w, the effects of pH on osmophilic yeasts might be the same as for non-osmophilic yeasts (38) because the pH range of HMFP (3.0–4.1) might not represent the optimum for growth.

Food preservation implies the placing of microorganisms in a hostile environment in order to inhibit their growth, shorten their survival, or cause their death (28). The feasible responses of microorganisms to such a hostile environment determines whether they grow or die. For the microbial stability of the developed HMFP, four or five hurdles or preservation factors were generally selected and include a blanching step, a water activity depression, lowering of pH, addition of chemical preservatives (sorbate or benzoate), and, in some cases, the addition of sulfite as an antibrowning agent. An important phenomenon in food preservation is the interference with the homeostasis of microorganisms (39). Homeostasis is the tendency to uniformity or stability in the normal status (internal environment) of organisms. If the homeostasis of microorganisms (internal equilibrium) is disturbed by the preservative factors (a_w, pH, additives, etc.) present in the food, the microorganisms will not multiply or may even die (28). Thus, food preservation is achieved by disturbing temporarily or permanently the homeostasis of the microorganisms in a food. Mechanisms of the main selected hurdles used in the preservation of HMFP on the microbial cell homeostasis are discussed next. Selected hurdles used in the preservation of particular foods have different mechanisms of action, and therefore microbial stability could be achieved with an intelligent combination of hurdles.

The water activity effect on the majority of microorganisms is related to the water status in their immediate environments because the microorganisms are metabolically active only in a range of water activities (40). Microorganisms do not contain water-impermeable barriers and tend to come rapidly into osmotic equilibrium with their surroundings. Therefore, in a food with a reduced a_w, the osmolality of the environment will be high and if the solutes present cannot penetrate the cell membrane, the microorganisms lose water until the equilibrium is re-established (40). The loss of water results in a reduced metabolic activity and, at least temporarily, growth ceases (40), affecting the homeostasis of the cell. Food pH is one of the principal factors that determine the survival and growth of microorganisms during food processing, storage, and distribution. Booth and Kroll (41) stated that microbial growth and viability depend on the cytoplasmatic pH homeostasis, which is regulated within relatively narrow limits. The microflora in a food is affected by the concentration of free hydrogen anions and by the concentration of undissociated weak acids (which is affected by the pH and the nature of the acid) (42). Weak acids cause leakage of hydrogen ions across the cell membrane, acidifying the cell interior and inhibiting transport mechanisms; growth inhibition can in principle be caused by inactivation of, or interference with, cell wall, cell membrane, metabolic enzymes, protein synthesis system, or genetic material (43). The cells of various microbial species present different tolerances toward internal acidification or anion accumulation, and their membranes have different permeability characteristics for lipophilic acids. Therefore, in a mixed initial flora, the acidity and pH of the food may determine the potential spoilage microorganisms. A low external pH leads to an initially high pH gradient; therefore, to achieve equilibration of the pH gradient, considerable amounts of the anion must be accumulated in the cell; the liberated protons exceed the capacity of the cell to maintain cytoplasmatic pH and microbial growth is inhibited (41). Besides the pH effects of sorbic and benzoic acids used as chemical preservatives, specific inhibition patterns are exhibited by the undissociated forms of these acids.

The water activity and pH hurdles, when used together with a blanching step that reduces the initial microbial count, disturb the homeostatic mechanisms of the microorganisms and assure microbial stability in fruits preserved by hurdle or combined-factors technology. For foods preserved by hurdle technology, different hurdles in a food may not exhibit an additive effect on stability but may act synergistically (4,27,44). The understanding of the influence of the hurdles on the homeostasis of microorganisms could lead to a "multi-target preservation" of foods (28), and therefore a minimal but effective preservation of foods.

The different preparation steps to which fresh fruits are subjected in the production process for HMFP will have a clear impact on the natural flora of fruits. Blanching is a critical control operation in the processing of HMFP. It is an early processing step for most vegetables and a few fruits, for which destruction of contaminating organisms is accomplished because the temperatures used are critical for some fungi (45). A decrease in the viable numbers is expected to occur during the a_w, pH, and preservatives equilibrium stage in which the blanched fruit pieces or purees are exposed to these factors, as well as during storage of the stabilized product. These phenomena have practical importance and had been called by Leistner (4) the "autosterilization" process of stable hurdle technology foods. This was observed by some researchers (46–48) in studies with high-moisture fruit products. The counts of a variety of bacteria, yeasts, and molds that survived blanching decreased below the detection limit during storage. Microorganisms in HMFP exert every possible repair mechanism to overcome the hostile environment; by doing this, the microorganisms completely use up their energy and die if they become "metabolically exhausted" (4).

Preserved papaya (a_w 0.98, 1000 ppm potassium sorbate, 150 ppm $NaHSO_3$, and pH 3.5) presented aerobic mesophilic, yeast, and mold counts < 10 CFU/g during 5 months of storage at 25°C (48). The microbiological analysis performed during 4 months of storage of pineapple slices preserved with a_w 0.97, 1000 ppm potassium sorbate, 150 ppm $NaHSO_3$, and pH 3.1 indicated that standard plate and yeast and mold counts remain below the level of detection for 30 days of storage at 27°C (49). In this study, also, no anaerobes were detected in the syrup or the fruit. The changes in the aerobic mesophilic plate count declined during storage and the viable population was almost lost after 20 days for banana puree preserved with a_w 0.97 adjusted with glucose, 100 ppm potassium sorbate, 400 ppm $NaHSO_3$, 250 ppm ascorbic acid, pH 3.5, and with a mild heat treatment; and for peach halves preserved with a_w 0.94 adjusted with glucose, 1000 ppm potassium sorbate, 150 ppm $NaHSO_3$ and pH 3.5. Banana puree and peach halves were stored at 25°C and 30°C, respectively (46,50). The microbiological stability obtained shows that it is more effective to use different factors or hurdles in small amounts in a food than only one factor in larger amounts because each factor used might hit different targets within the microbial cell and thus act synergistically (28). However, the response to the stress factors of certain key microorganisms has to be studied using different approaches. Some basic information is required to establish the types and numbers of microorganisms to be used in a microbial challenge test (31). The choice of the proper organism to challenge the product is very important, and isolates from the fruit product under test would be used when possible (51). The microbial challenge test can be used as a model to simulate product behavior during processing, distribution, and storage. For the selection of the microorganism(s) to be inoculated, the results obtained in studies carried out in vitro are a useful key: Cerrutti et al. (52) reported different combinations of a_w, pH, and potassium sorbate and/or sodium bisulfite that inhibit yeast growth in laboratory media: a_w 0.97, pH 4.0, and 500 ppm of potassium sorbate; or with the same a_w, but with pH 3.0 and 200 ppm of the preservative. Tapia de Daza and Aguilar (53) reported growth inhibition in laboratory media of *Zygosaccharomyces rouxii*, isolated from an intermediate moisture papaya, at a_w 0.95 with pH 4.0, 500 ppm of potassium sorbate, and 100 ppm of sodium bisulfite. López-Malo et al. (54) reported the individual and combined effects of a_w (0.99, 0.97, and 0.95), pH (5.5, 4.5, and 3.5) and potassium sorbate concentration on the extension of the lag phase and the rate of radial growth of *Aspergillus flavus*, *A. ochraceus*, and *Penicillium* sp. in a potato dextrose agar system, and reported several inhibitory combinations of the studied factors. As water activity decreases, the pH and preservative concentration needed to suppress growth decrease.

Tablante and Tapia de Daza (55) performed a challenge study in intermediate-moisture papaya (a_w 0.89, pH 4.0, 1000 ppm of potassium sorbate, and 150 ppm of sodium bisulfite) inoculating *Z. rouxii, Saccharomyces cerevisiae, Aspergillus ochraceus,* and *Staphylococcus aureus*. The osmotolerant yeast *Z. rouxii* was the only microorganism that proliferated with no difficulty in the product after an initial decline; the rest of the organisms are inhibited during the first week of storage. *Z. rouxii* was inoculated in two different high-moisture papaya products, one with a_w 0.98, pH 3.5, 150 ppm sodium bisulfite, 1000 potassium sorbate (56) and the other with a_w 0.96, pH 3.5, 150 ppm sodium bisulfite, 1000 ppm potassium sorbate (57); in both cases the selected hurdles reduced the initial count of the yeast but the surviving cells were viable even after 3 months of storage. In these cases, the conditions selected for the studies were sufficient to inhibit yeast growth, and although the factors used were not enough to inactivate the yeast, they were sufficient to maintain microbiologically sound food products.

According to the behavior of the native flora and the results obtained with microbiological challenge tests, the processed fruits were microbiologically stable in non-refrigerated storage for 3 to 8 months. Even so, sanitation is an essential prerequisite for production, and contamination can always be minimized by employing proper sanitary and handling techniques. The design of a proper Hazard Analysis Critical Control Point (HACCP) program to assure safety and overall quality of HMFP is necessary, and it should not be a complex task; since the type of products, due to intrinsic and processing conditions, should not represent significant risks of pathogenic hazards, while spoilage potential can be successfully predicted and controlled (31).

IV. QUALITY FACTORS

Consumers of fruits are putting increased emphasis not only on convenience but on quality and fresh-like characteristics (36). These consumption patterns have led to the development of techniques of food preservation that produce minimally processed fruits with improved quality and extended shelf-life. The combined-methods technology, as a preservation technique, has as its major goal microbial stability, but even food that is microbiologically sound may deteriorate. The basis of most food preservation techniques is inhibition of the growth of food-spoilage and pathogenic microorganisms, but preserving other food attributes is an additional concern (40). There is an increasing demand for quality attributes in minimally and traditionally processed fruits (58). Argaiz et al. (37) mentioned that fruit processed by combined-methods technology must have a consistent texture, fresh appearance, and acceptable color.

Alzamora et al. (36) and Argaiz et al. (37) reported that in addition to microbial spoilage, physicochemical changes can occur during processing and storage of fruits preserved with hurdle technology. To extend the high quality of HMFP, the proper maturity at harvest is a determining factor; this leads to placing more emphasis on visual characteristics (37) and sacrificing others, like flavor. Some steps or hurdles applied during the process for stabilization of fruit by combined methods affect their physicochemical and sensory characteristics. However, some of the hurdles help to maintain food color, texture, and sensory characteristics during storage.

For the preservation of fruit, two hurdles are applied to extend shelf-life—blanching and the addition of sulfite. The slight thermal treatment applied as the blanching step inactivates enzymes and prevents deterioration reactions and the sulfiting agents (added in some cases) inhibit or delay enzymatic and nonenzymatic browning reactions.

Other factors may influence the characteristics of HMFP, such as low pH and high water activity, which exert a determinant role in many chemical and enzymatic reactions that take place during nonrefrigerated storage. Leung (59) established that water may influence chemical reactivity in different ways. Water may act as a reactant, as in sucrose hydrolysis. Water as a solvent exerts a dilution effect. Water also may change the mobility of or interact with the reactants. Water accelerates the browning reactions, imparting mobility to the substrates, or may decrease the reaction rate by dilution of the reactants (59,60). Browning rates generally increase with increasing a_w, reaching a maximum between a_w of 0.40 and 0.80. Leung (59) reported that in intermediate- or high-moisture a_w range, chemical reaction rates may decrease with increasing a_w because of dilution effects.

Table 2 presents the sensory characteristics of papaya and pineapple preserved by combined methods. The scores are the mean values obtained with an untrained panel of 50 judges and represent the initial sensory characteristics for a shelf-life study. Scores correspond to products with good acceptability (overall acceptability around 7, "like moderately"). Similar results were obtained for pineapple slices by Alzamora et al. (49) and for peach halves by Argaiz and López-Malo (61). In general, the application of hurdle technology results in products with good sensory characteristics. The combined-methods technology mainly consists of placing the blanched fruit pieces into a solution of sucrose, acids, and additives in such concentrations

TABLE 2 Sensory Characteristics of
Papaya and Pineapple Preserved by
Combined Methods

Attribute [a]	Papaya	Pineapple
Odor	6.48	6.36
Color	7.21	7.09
Texture	7.44	7.36
Flavor	7.33	6.98
Overall acceptability	7.25	7.21

[a] Results obtained by untrained panels (50
members) using a nine-point hedonic scale (9,
like extremely; 1, dislike extremely)
Source: From Refs 48 and 71.

that after the stabilization time, the desired water activity, pH, and additive
concentrations are obtained; thus several changes in fruit composition take
place during this stabilization process. The major changes are the result of an
osmotic concentration process.

Table 3 presents the initial characteristics of fresh, blanched, and preserved (immediately after processing) papaya, pineapple, and peaches. The
fruits were processed with a sucrose-additives syrup of such composition that
at equilibrium, a water activity of 0.98 was obtained as well as potassium sorbate and sodium bisulfite concentrations of 1000 and 150 ppm, respectively.
The syrup pH was reduced with citric acid for the papaya and pineapple
slices, reaching a final value in the fruits of 3.5 and 3.7, respectively. Peach
halves syrup was treated with a mixture of citric and phosphoric acids to at-
tain pH 3.7 in the fruits. Moisture content of the fruits increased slightly after
blanching, reflecting the leaching of soluble solids. The preservation method
reduced the moisture content because of the water and solute exchange between fruit and syrup during osmotic concentration. For pineapple and
peaches, the equilibrium water activity was near the desired values. The equilibrium water activity for the papaya was lower than desired, which was calculated using sucrose as a base. The reducing sugar content of papaya was
predominant, reflecting sucrose hydrolysis. Sucrose chemical and/or enzymatic hydrolysis can occur and therefore modify the a_w of the preserved fruit
because of the great capacity of fructose and glucose for lowering a_w (62), increasing the effect of the a_w hurdle on microbial growth (36,48). However, su-
cose hydrolysis favors non-enzymatic browning with related changes in color

TABLE 3   Initial Characteristics of Fresh, Blanched, and Preserved Fruits

	Fresh	Blanched	Preserved
Papaya			
Moisture content (%)	91.27	92.02	72.62
pH	5.4	5.6	3.5
Brix degree	7.7	7.3	27.2
Reducing sugars	7.7	7.2	23.5
Ascorbic acid	46.4	33.1	23.1
Water activity	0.996	0.996	0.963
Color:			
Lightness	44.2	47.62	59.37
Hue	64.48	65.44	73.35
Chroma	30.5	32.98	38.81
Texture (kg)	6.68	5.36	5.41
Pineapple			
Moisture content (%)	87.57	88.2	70.68
pH	2.97	3.13	3.69
Brix degree	10.7	10.2	25.9
Reducing sugars	4.7	4.1	2.8
Water activity	0.994	0.986	0.978
Color:			
Lightness	79.87	73.14	69.26
Hue	−84.1	−83.84	−84.21
Chroma	34.92	37.31	33.87
Texture (kg)	6.21	6.09	5.83
Peach			
Moisture content (%)	86.6	89.4	77.3
pH	3.75	3.79	3.71
Brix degree	6.5	5.6	25.5
Reducing sugars	1.4	2.0	1.8
Water activity	0.995	0.997	0.979
Color:			
Lightness	49.48	50.16	53.31
Hue	70.53	71.08	76.65
Chroma	29.36	30.25	33.65

Source: From Refs 48, 61, and 71.

(63). Montes de Oca et al. (64) reported that the presence of fruit juices increases the rate of sucrose hydrolysis, probably because of the catalytic effect of the fruit organic acids, and reported that the presence of fruit juices in sucrose syrups affects their a_w. Complete sucrose hydrolysis in plum juice after 20 days of storage at 37°C was observed (64).

Blanching did not change considerably the lightness, hue, and chroma values of papaya, pineapple, and peaches (Table 3). The color parameters changed for preserved fruits, reflecting the sugar uptake during osmotic concentration. The blanching process produced a softening of the fruit tissue, which in general was not impaired during the osmotic treatment. The mean texture values for fresh and blanched fruits did not differ significantly. Monsalve-González et al. (65) reported a softening of unblanched apple slices during the osmotic adjustment of water activity. Maltini et al. (66) reported that during an osmotic treatment to adjust a_w of fruit pieces, the solid gain is greatly enhanced in previously blanched fruits. The soluble solids intake had only a limited effect on the texture of the fruit. Fruit texture depends primarily on the insoluble matter and on the moisture content rather than on the soluble solids and a_w. Therefore, a low a_w may be achieved while maintaining an acceptable texture.

Changes in the color of processed fruits were associated with enzymatic and nonenzymatic browning reactions, as well as with natural pigments that may degrade or be lost by leaching (36). Sulfites may be added to foods to inhibit enzymes, growth of microorganisms, or nonenzymatic browning. (67). Reactions between sulfites and carbonyl compounds are important in preventing nonenzymatic browning. Many foods develop a slight to medium brown color during processing or storage that is generally considered unacceptable. The effectiveness of sulfite in preventing nonenzymatic browning depends on the number of carbonyl groups in the intermediates that are formed, and on the free carbonyl compounds present.

Many fresh foods contain active enzymes, which can chemically or physically change foods. Major visual changes in foods result from the enzymatic and/or non-enzymatic development of brown pigmented substances (68). Enzymes also cause flavor and textural changes. Texture can be affected by enzymatic, physiological, and/or physicochemical factors (69). Sulfite is often added to foods to inhibit enzyme action, especially to prevent enzymatic browning caused often by polyphenoloxidases (67). Adding sulfites to fruits preserves the organoleptic quality of products, predominantly by delaying the onset of deterioration during storage (70).

Compared with some other widely used preservatives, sulfites are relatively unstable in foods and therefore may lose effectiveness during storage.

Initial sulfite concentrations in a food may often decrease during processing and storage. For example, the boiling of an acid food like sulfited strawberries during the manufacture of jam resulted in sulfite losses as high as 95%. Such losses must be taken into account because of legal constraints and for designing processes and storage conditions (70). Argaiz et al. (37) reported that for high moisture fruit products, the loss of sulfites depends on storage temperature and packaging material.

Alzamora et al. (49) observed, during storage, slight discoloration of pineapple slices preserved by combined methods, probably due to the bleaching of the fruit pigments produced by sulfur dioxide. The L value for papaya preserved with a_w 0.98 adjusted with sucrose, 1000 ppm potassium sorbate, 150 ppm $NaHSO_3$, and pH 3.5 decreased between 10% and 20% during storage at 5°C and 25°C respectively, probably as a result of nonenzymatic browning reactions due to the high reducing sugar content (48). Added sulfites were rapidly lost from the fruit, with a retention after 5 months' storage of approximately 63%. Hue angle and chroma remained constant throughout storage at both temperatures. López-Malo et al. (71) reported that for pineapple slices preserved by combined methods (a_w 0.98 adjusted with sucrose, 1000 ppm potassium sorbate, 150 ppm $NaHSO_3$, pH 3.7,) the color parameters hue and chroma changed slightly during the first 15 days of storage at 5° and 25°C, and remained constant after 3 months at both temperatures. In this case, after 3 months the total sulfite retention was 51% and 38% for 5° and 25°C respectively. The 3.7 pH of the pineapple slices favors the formation of bisulfite anion, which is less volatile than the sulfite anion. Alzamora et al. (49) reported a loss of approximately 50% of the initial sulfite in pineapple slices immersed in glucose syrup during 4 month of storage at 27°C. Argaiz (72) and Argaiz et al. (73,74) reported a rapid loss of 60% of the total sulfite in fruits immersed in sucrose syrups and stored in glass jars at 35°C. Similar results in color retention were observed by Argaiz et al. (74) for peach halves preserved by combined methods during 3 months of storage at 35°C. Monsalve-González et al. (65) indicated that the L and a color values of unblanched apple slices preserved by combined methods (a_w 0.96–0.97 adjusted with sucrose, pH 3.9–4.2, 0.15% w/v sorbic acid) almost remained unchanged during 75 days' storage at 25°C. Apple slices were immersed in 1000 ppm sodium sulfite or 200 ppm 4-hexylresorcinol solutions for 1 or 10 h and stored at 35°C. The sulfiting agent was more effective than 4-hexylresorcinol in preventing either enzymatic or nonenzymatic browning. During storage at 45°C, the treated apples browned regardless of the antibrowning treatment.

The color changes reported in Table 3 did not significantly affect the sensory acceptability of papaya, pineapple, and peaches preserved by combined methods (Table 4). The sensory evaluation scores obtained correspond to products with good acceptability. These results revealed that fruit preserved with the combined-method technology are acceptable after several months of storage. Color measurements indicated that stability during processing and storage is dependent on the type of fruit, the preservation method, and the storage temperature (36). The onset and extension of browning during storage can be correlated with the chemical and/or physical depletion of antibrowning agents in the fruit tissue (48,50,75,76). Sulfur dioxide is depleted in stored fruits, decreasing its effectiveness as a hurdle to microbial growth and/or nonenzymatic browning. López-Malo et al. (77) reported a direct relationship between sulfite loss and the onset of browning in apple slices preserved by combined methods, packaged in polyethylene bags and stored at selected temperatures. At 25°C, browning initiated after 31 days of storage, for 35°C after 24 d. Only 11 days of storage were needed to initiate browning in apple slices stored at 45°C.

Sulfites are good inhibitors of nonenzymatic and enzymatic browning, but their use is being prohibited because certain groups of people, especially asthmatics, are adversely affected (78). The use of sulfites for fruits and vegetables has been banned since 1986 (except for potatoes, wines and certain

TABLE 4 Sensory Evaluation of Fruits Preserved by Combined Methods after 2–5 Months of Storage

Fruit and conditions of storage	Attribute[a]				
	Flavor acceptability	Odor	Texture	Color	Overall
Papaya slices					
3 months 35°C	7.12	6.24	7.22	6.46	6.93
5 months 25°C	6.65	5.83	7.07	6.91	6.73
Pineapple slices					
3 months 35°C	7.70	6.73	8.07	7.10	7.63
3 months 25°C	6.96	6.91	7.33	6.87	7.02
4 months 27°C	7.00	NE[b]	NE[b]	NE[b]	NE[b]
Peach halves					
2.5 months 35°C	NE[b]	NE[b]	NE[b]	NE[b]	6.87

[a] Results obtained by untrained panels (30 to 50 members) using a nine point hedonic scale (9, like extremely; 1, dislike extremely).
[b] Not evaluated.
Source: From Refs 47, 48, 49, 71, and 74.

dried fruits) by the FDA (79). It is necessary to find substitutes for sulfites as browning inhibitors.

To reduce or inhibit the enzymatic browning of apples, thermal treatments, oxygen removal, the use of ascorbic acid, erythorbic acid and their derivatives (80–85), cysteine, sequestrants (e.g., EDTA, pyrophosphates), citric acid, inhibitors such as 4-hexylresorcinol (75), calcium (80,81), aliphatic alcohols, some enzymes (86,87), and growth regulators of fruits, like gibberellic acid and etephon, have been suggested (88). Some of these substances can be combined to inhibit browning. Currently there exist commercial combinations of inhibitors such as Snow Fresh™(89), Color Fresh™, and Freshway™ (87). Most of the mixtures are a combination of ascorbic and citric acids.

Monsalve-González et al. (75) reported 4-hexylresorcinol–ascorbic acid combinations effectively controlled browning during storage of apple slices preserved by combined methods. Zaragoza et al. (90) reported that treatments with 4-hexylresorcinol and sulfites (150 ppm) provided protection against browning of apple preserved by combined methods during the first 10 days of storage. A mixture of calcium chloride, sodium acid pyrophosphate, and ascorbic acid were effective in protecting against browning during 80 days of storage. Argaiz et al. (37) reported that a mixture of calcium chloride, ascorbic acid, and sodium acid pyrophosphate, in addition to reduced a_w and pH, maintained color in apple slices during 2 months of storage at 25°C. The osmotic treatments applied in combined methods to stabilize the fruits partially inhibit enzymatic browning by reducing diffusion of oxygen to the fruit. However, browning is still a problem that reduces acceptance in fruit products with a high inherent polyphenoloxidase activity (91).

Consumers like food to be moist and tender and associate adjectives like moist, juicy, tender, and chewy with desirable textural attributes (92). A high a_w level in a food product may have a great impact on consumer demands. The texture of the parenchyma of fruits is governed by the size and shape of the cells, the ratio of the cytoplasm to the vacuoles, the volume of intercellular spaces, and the thickness and molecular composition of the cell wall and the osmotic pressure and type of solutes present (93). Therefore, each fruit exhibits different responses to blanching, osmotic treatments, and storage conditions, although the combined methods do not greatly modify the texture of the fresh fruit. The effect of a_w on food texture is specific to the kind of food under consideration (92). To obtain desirable textures it is usually necessary that foods have a high moisture content.

Alzamora et al. (49) reported pineapple texture was not affected by blanching, osmotic treatment (a_w 0.97, glucose), and storage at 27°C. The shear force value was constant during 4 months of storage, although pineapples became somewhat more flexible. López-Malo et al. (48) reported compression force values for papaya were constant for 150 days of storage at 5°C. The firmness of preserved papaya stored at 25°C decreased slightly. Because limited heat treatment was given to the fruit during vapor blanching (30 s at 100°C), the difference in texture between 5° and 25°C storage was attributed to higher activity of papaya pectolytic enzymes at 25°C. However, variations in texture after blanching procedure and storage at 5° or 25°C were insufficient to cause an important sensory change.

Strawberries were softened by blanching. The textural quality of water-blanched strawberries diminished significantly with storage, reaching very low values (94). The addition of 0.1% w/w calcium lactate (maximum concentration admissible for sensory acceptability) during the osmotic concentration stage improved textural characteristics of vapor-blanched and sugar-treated strawberries but was ineffective in preventing softening of water-blanched and sugar-treated strawberries. Monsalve-González et al. (65) reported that treatments with calcium chloride minimize the apple softening that results from osmotic treatment to reduce the water activity to 0.96.

Bourne (95) reported the texture profile analysis parameters of fresh apples and those equilibrated to different a_w levels. The fresh apples (a_w 0.99) were rigid and exhibited a fracturability peak and low cohesiveness. In contrast, the apples equilibrated to a_w 0.85 exhibited no fracturability and high deformability. The change in a_w from 0.99 in fresh apples to 0.85 (due to moisture reduction) causes a complete loss of the turgor pressure characteristic of fresh apples, resulting in a total loss of fracturability, an increase in deformability, and a moderate increase in cohesiveness (95). Monsalve-González et al. (65) reported a softening of unblanched apples that occurs at the early stages of an osmotic treatment to decrease a_w to 0.96; the softening was limited to the edges of the apple tissue where the sugar had penetrated. A negative correlation was observed between sucrose uptake and loss of texture, although factors such as enzymatic action, calcium leaching from the cell wall, solubilization of protopectin, loss of cell turgor, and/or degradation of the middle lamellae could also contribute to texture changes. Beveridge and Weintraub (96) reported that the force required to shear blanched apple slices stabilized at a_w between 0.70 and 0.77 was similar to that for sulfited or untreated ones, but the force needed to shear the blanched slices with a_w below 0.57 was higher than that for the sulfited or untreated apples. Blanching apple slices to control enzymatic browning is a useful technique for apple products with a reduced a_w, but it is detrimental to texture quality when the apple tissue is to be stored at a_w below 0.60.

V. HURDLE TECHNOLOGY AND MINIMALLY PROCESSED FOODS

The term *minimally processed* covers a wide range of methods and technologies for preserving foods that induce a minimum change in the fresh-like quality characteristics of the food (97). Nguyen-the and Carlin (98) defined minimally processed fruits and vegetables as fresh, raw fruits or vegetables processed to supply a ready-to-eat or ready-to-use food product. The operations traditionally used in the minimal processing of foods are trimming, peeling or cutting, washing, disinfecting, and packaging. A refrigerated storage is necessary to assure a shelf life of several days. These minimally processed food products can be used as ingredients in cooked dishes or are consumed raw (98).

Minimally processed foods are the response of the food industry in the direction of the consumption patterns. The demand for "reduced" and "light" products represents a major consumer trend aimed at lowering the caloric content or the fat, sugar or, salt contents of the foods; some additives tend to be removed and replaced by natural ingredients, in order to make more natural or more fresh-like food products (97). Fresh-like products are more perishable. Much emphasis has to be put on product safety and quality. The long shelf life of food products is not an important selling point anymore. Preservation techniques that prolong storage stability and do not have a detrimental effect on quality are the major goal. There is a strong need from food processors for new or improved mild preservation methods that allow the production of fresh-like, but stable and safe foods (29).

Hurdle technology uses the intelligent combination of traditional or novel preservation techniques to assure that the microorganisms in the food product are not able to grow (4). Using the hurdle concept, several high-moisture fruit products were developed and their stability was established (31–37). Huxsoll and Bolin (99) mentioned that these products may be called partially processed instead of minimally processed because of the additional processes included; but because the fruit products are intended to maintain fresh-like appearance and quality it is appropriate to consider them minimally processed (31). The stability of fruits minimally processed by hurdle technology was evaluated by determining the native microbial load of yeasts, molds, and aerobic and anaerobic mesophilic and thermophilic microorganisms of fruits and syrups in different stages of the preservation process and during storage. Heating during blanching significantly reduces yeast, mold and aerobic microorganisms counts. The number of microorganisms is lowered still more during the equilibration stage, probably because of heat's preservative-sensitizing effect on the survivors. As an example, consider the microbiological condition of the raw and blanched fruit and the microbiological status of the finished minimally processed pineapple and papaya of Alzamora et al. (49) and López-Malo et al. (48). Blanching reduced the microbial load from 60% to 99%, bringing the counts after stabilization very low or under the level of detection.

The high-moisture fruit products were developed taking into consideration microbial stability and an extended shelf life of at least 3 months without refrigeration. However, some fruit products have intense sweet and/or acid flavors and include the use of additives like sulfite, which is of much concern to consumers and in some countries is banned for use with some fruit products (37). For the development of novel minimally processed fruit products under the hurdle concept, the chemical overload must be reduced and the combination of factors may include other hurdles such as refrigeration, slight thermal or nonthermal treatments, and the use of natural antimicrobials. Argaiz et al. (37) recommends further research to develop predictive models for microbial growth as well as deteriorative reactions. The predictive models can be used to better select the hurdles and their levels to assure the stability and quality of the minimally processed fruit products.

VI. FUTURE TRENDS

Studies that consider the possible presence of heat-resistant molds that traditionally have been implicated in spoilage outbreaks of fruit juices and other heat-processed fruit-based products are needed. The examination for *Byssochlamys* species, *Talaromyces flavus*, and *Neosartoria fisheri* can be helpful (100). Inhibition of sorbate-resistant yeasts and molds has to be studied. *Zygosaccharomyces bailii*, another organism of concern in HMFP, exhibits considerable resistance to heat, low pH, reduced a_w, and sorbic and benzoic acids (101,102). Preconditioning to sorbate and/or solutes can affect *Z. Bailii* sorbate resistance (103); this fact has to be addressed when solving the technological problem represented by the remaining spent syrup after the stabilization of fruit pieces. The reuse of the syrup may become a risk in relation to a buildup of spoilage yeasts, and therefore syrup reuse in the HMFP processes should only be recommended after a pasteurization process.

The utilization of chemical or synthetic preservatives to prevent the growth of food poisoning and food spoilage microorganisms is being challenged by consumers because of the safety of several food additives is questionable (36). Antimicrobial systems naturally present in plants are extremely appealing to the public in the context of "natural preservatives." Plant parts and extracts from several types of plants used as flavoring agents are known to possess antimicrobial activity and seem to be suitable for fruit products (104–106). Recent reports of in vitro and in-food antimicrobial activities of spice essential oils indicate that the concentrations necessary to inhibit microbial growth are considerably higher in foods than in culture media, significantly impairing the flavor of foods. However, the effectiveness of spice essential oils can be enhanced when they are used in combination with other stress factors or hurdles (107). The use of natural preservatives as one of the hurdles has been reported by Alzamora et al. (36) and López-Malo et al. (71) for the stabilization of fruit purees against fungi growth with vanillin (4-hydroxy-3-methoxybenzaldehyde). Cerrutti and Alzamora (108) reported that addition of 2000 ppm vanillin had an important inhibitory effect on *Saccharomyces cerevisiae*, *Zygosaccharomyces bailii*, *Z. rouxii* and *Debaryomyces hansenii* in Sabouraud broth at pH 4.0 and apple puree at pH 3.5, a_w 0.99 or 0.95, for 40 days' storage. Also, López-Malo et al. (109) reported that 1500 ppm of vanillin was inhibitory for *Aspergillus flavus*, *A. ochraceus*, and *A. parasiticus* in mango-, papaya-, pineapple-, apple-, and banana-based agars (a_w 0.98 and pH 3.5). *A. niger* growth inhibition was achieved with 2000 ppm vanillin. Results suggest that vanillin may be a promising additive for inhibiting fungi in fruits and fruit products, depending on the composition.

Mano-thermo-sonication represents an interesting combination of low pressure (0.3 MPa), mild heat treatment, and ultrasonic wave treatment that seems effective for inactivation of microorganisms (110). The inactivation of peroxidase, lipoxygenase, and polyphenol oxidase by mano-thermo-sonication was studied by López et al. (6). A synergistic effect that substantially reduced enzyme resistance and the heat treatment required for inactivation was observed. The combined treatment could be used—in milk, juices, and other drinks—to solve problems caused by thermostable enzymes. Research is in progress to clarify the mechanisms responsible for the observed enzyme inactivation under this new technique.

Osmodehydrofreezing, a combined process in which osmotic dehydration is followed by air dehydration and freezing, was proposed to prepare reduced-moisture vegetable ingredients free of preservatives, with a natural and agreeable flavor, color, and texture, and with functional properties suitable for different applications (111). Osmotic concentration followed by air drying using the hurdle technology was also proposed as an alternative to traditional drying (112–114). The influence of vacuum treatment on mass transfer during osmotic dehydration was recently studied (115,116). This method takes advantage of the porous microstructure of some fruits, increasing water transfer rate.

The overview presented in this chapter serves to highlight the fact that there is a growing number of improved technologies, and new ones that are being researched or are in the early stages of application. The use of the hurdle technology concept could start an explosion of investigation in the area of new processes and products, especially the exploration of the combination of traditional factors with the nonthermal methods covered in this book. Other novel interactions can be the combination of new antimicrobials, edible films, or controlled-atmosphere packaging, just to mention a few of the many possibilities to investigate. The technical and economical feasibility as well as the safety and total quality management of the new technologies are among the many challenges facing food scientists today.

REFERENCES

1. G. W. Gould and M. V. Jones, Combination and synergistic effects, *Mechanisms of Action of Food Preservation Procedures* (G.W. Gould, ed.), Elsevier, London, 1989, pp. 401-422.
2. L. Leistner and W. Rödel, The stability of intermediate moisture foods with respect to micro-organisms, *Intermediate Moisture Foods* (R. Davies, G.G. Birch, and K.J. Parker, eds.), Applied Science Publishers, London, 1976, pp. 120-137.
3. L. Leistner, Hurdle technology applied to meat products of the shelf stable products and intermediate moisture food type, *Properties of Water in Foods in Relation to Quality and Stability* (D. Simatos and J. L. Multon, eds.), Martinus Nijhoff Publishers, Boston, 1985, pp. 309-329.
4. L. Leistner, Principles and applications of hurdle technology, *New Methods of Food Preservation* (G.W. Gould, ed.), Blackie Academic and Professional, Glasgow, 1995, pp.1-21.
5. C. Grijspaardtvink, Food preservation by hurdle technology. *Food Technol.* 48(12): 28 (1994).
6. P. López, F. J. Sala, J. L. de la Fuente, S. Condón, J. Raso, and J. Burgos, Inactivation of peroxidase, lipoxygenase, and polyphenol oxidase by manothermosonication. *J. Agric. Food Chem.* 42(2): 252-256 (1994).
7. H. Vega-Mercado, U. R. Pothakamury, F. J. Chang, G. V. Barbosa-Cánovas, and B.G. Swanson, Inactivation of *Escherichia coli* by combining pH, ionic strength, and pulsed electric fields. *Food Research International*, 29: 117-121 (1996).
8. X. Liu, A. E. Yousef, and G. W. Chism, Inactivation of *Escherichia coli* O157:H7 by the combination of antimicrobial organic acids and pulsed electric fields, *1995 IFT Annual Meeting: Book of Abstracts*, 1995, p. 29.
9. M. V. Simpson, G. V. Barbosa-Cánovas, and B. G. Swanson, The combined inhibitory effect of lysozyme and high voltage pulsed electric fields on the

growth of *Bacillus subtilis* spores, *1995 IFT Annual Meeting: Book of Abstracts*, 1995, p. 267.

10. D. Knorr, High pressure effects on plant derived foods, *High Pressure Processing of Foods* (D. A. Ledward, D. E. Johnston, R. G. Earnshaw, and A. P. M. Hasting, eds.), Nottingham University Press, Nottingham, 1995, pp. 123-136.

11. M. F. Patterson, M. Quinn, R. Simpson, and A. Gilmour. Effects of high pressure on vegetative pathogens, *High Pressure Processing of Foods* (D. A. Ledward, D. E. Johnston, R. G. Earnshaw, and A. P. M. Hasting, eds.), Nottingham University Press, Nottingham, 1995, pp. 47-64.

12. B. Mertens and G. Deplace, Engineering aspects of high pressure technology in the food industry. *Food Technol.* 47(6): 164-169 (1993).

13. C. M. Roberts and D. G. Hoover, Tolerance of *Bacillus coagulans* 7050 spores to combinations of high hydrostatic pressure, heat, acidity, and nisin, *1995 IFT Annual Meeting: Book of Abstracts*, 1995, p. 268.

14. E. Palou, A. López-Malo, G. V. Barbosa-Cánovas, J. Welti, and B. G. Swanson, High hydrostatic pressure as a hurdle for *Zygosaccharomyces bailii* inactivation, *J. Food Sci.*, 62(4) (1997).

15. L. Popper and D. Knorr, Applications of high-pressure homogenization for food preservation. *Food Technol.* 44(7): 84-89 (1990).

16. A. M. Papineau and T. Schmersal. 1987. Unpublished data, University of Delaware, Newark. Cited by: D. G. Hoover, C. Metrick, A. M. Papineau, D. F. Farkas, and D. Knorr, Biological effects of high hydrostatic pressure on food microorganisms. *Food Technol.* 43(3): 99-107 (1989).

17. D. Knorr, Hydrostatic pressure treatment of food: microbiology, *New Methods of Food Preservation* (G.W. Gould, ed.), Blackie Academic and Professional, New York, 1995, pp. 159-175.

18. K. J. A. Hauben, E. Y. Wuytack, C. F. Soontjens, and C. W. Michiels, High-pressure transient sensitization of *Escherichia coli* to lysozyme and nisin by disruption of outer-membrane permeability. *J. Food Prot.* 59: 350-355 (1996).

19. Y. J. Crawford, E. A. Murano, D. G. Olson, and K. Shenoy, Use of high hydrostatic pressure and irradiation to eliminate *Clostridium sporogenes* spores in chicken breast. *J. Food Prot.* 59(7): 711-715 (1996).

20. R. G. Earnshaw, Kinetics of high pressure inactivation of microorganisms, *High Pressure Processing of Foods*. (D. A. Ledward, D. E. Johnston, R. G. Earnshaw, and A. P. M. Hasting, eds.), Nottingham University Press, Nottingham, 1995, pp. 37-46.

21. B. Tauscher, Pasteurization of food by hydrostatic pressure: chemical aspects. *Zeitschrift für Lebensmittel-Untersuchung und-Forschung*, 200: 3-13 (1995).

22. E. Palou, A. López-Malo, G. V. Barbosa-Cánovas, J. Welti, and B. G. Swanson, Combined effect of water activity on high hydrostatic pressure inhibition of *Zygosaccharomyces bailii*. *Lett. Appl. Microbiol.*, 24: 417-420 (1997).

23. J. C. Cheftel, High-pressure, microbial inactivation and food preservation. *Food Sci. Technol. Int.* 1: 75-90 (1995).

Nonthermal Preservation of Foods

24. J. Chirife and G. J. Favetto, Some physico-chemical basis of food preservation by combined methods. *Food Research International* 25: 389-396 (1992).

25. M. Loncin, Basic principles of moisture equilibria, *Freeze Drying and Advanced Food Technology* (S. A. Goldblith, L. Rey, and W. W. Rothmayr, eds.), Academic Press, London, 1976, pp. 559-618.

26. L. Leistner, Shelf-stable products and intermediate moisture foods based on meat, *Water Activity: Theory and Applications to Food.* (L. B. Rockland and L. R. Beuchat eds.), Marcel Dekker, New York, 1987, pp. 295-328.

27. L. Leistner, Food presevation by combined methods. *Food Research International* 25: 151-158 (1992).

28. L. Leistner, Use of hurdle technology in food processing: recent advances, *Food Preservation by Moisture Control. Fundamentals and Applications* (G. V. Barbosa-Cánovas and J. Welti-Chanes, eds.), Technomic, Lancaster, PA, 1995, pp. 377-396.

29. L. Leistner and L. G. M. Gorris, Food Preservation by Hurdle Technology. *Trends in Food Science and Technology* 6(2): 41-46 (1995).

30. L. Leistner, W. Rödel, and K. Krispien, Microbiology of meat and meat products in high-and intermediate moisture ranges, *Water Activity: Influences on Food Quality* (L. B. Rockland and G. F. Stewart, eds.), Academic Press, New York, 1981, pp. 855-916.

31. M. S. Tapia de Daza, A. Argaiz, A. López-Malo, and R. V. Díaz, Microbial stability assessment in high and intermediate moisture foods: special emphasis on fruit products, *Food Preservation by Moisture Control. Fundamentals and Applications* (G. V. Barbosa-Cánovas and J. Welti-Chanes, eds.), Technomic, Lancaster, PA, 1995, pp. 575-602.

32. K. S. Jayaraman and D. K. das Gupta, Development and storage stability of intermediate moisture carrot. *J. Food Sci.* 43: 1880-1881 (1978).

33. A. Levy, S. Gagel, and B. Juven, Intermediate moisture tropical fruit products for developing countries. I. Technological Data on Papaya. *J. Food Technol.* 18: 667-685 (1983).

34. K. S. Jayaraman, Development of intermediate moisture tropical fruit and vegetable products—technological problems and prospects, *Food Preservation by Moisture Control* (C. C. Seow, ed.), Elsevier, London, 1988, pp.175-198.

35. J. Welti, M. S. Tapia de Daza, J. M. Aguilera, J. Chirife, E. Parada, A. López Malo, L. C. López, and P. Corte, Classification of intermediate moisture foods consumed in Ibero-America, *Revista Española de Ciencia y Tecnologa de Alimentos* 34: 53-63 (1994).

36. S. M. Alzamora, P. Cerrutti, S. Guerrero, and A. López-Malo, Minimally processed fruits by combined methods, *Food Preservation by Moisture Control. Fundamentals and Applications* (G. V. Barbosa-Cánovas and J. Welti-Chanes, eds.), Technomic, Lancaster, PA, 1995, pp. 463-492.

37. A. Argaiz, A. López-Malo, and J. Welti, Considerations for the development and the stability of high moisture fruit products during storage, *Food Preserva-*

tion by Moisture Control. Fundamentals and Applications (G. V. Barbosa-Cánovas and J. Welti-Chanes, eds.), Technomic, Lancaster, PA, 1995, pp. 729-760.

38. R. H. Tilbury, The microbial stability of intermediate moisture foods with respect to yeasts, *Intermediate Moisture Foods*, (R. Davies, G.G. Birch, and K.J. Parker, eds.), Applied Science Publishers, London, 1976, pp. 138-165.

39. G. W. Gould, Interference with homeostasis—Food, *Homeostatic Mechanisms in Micro-organisms* (R. Whittenbury, G. W. Gould, J. G. Banks, and R. G. Board, eds.), Bath University Press, Bath, 1988, pp. 220-228.

40. G. W. Gould, Drying, raised osmotic pressure and low water activity, *Mechanisms of Action of Food Preservation Procedures* (G. W. Gould, ed.), Elsevier, London, 1989, pp. 97-118.

41. I. Booth and R. G. Kroll, The preservation of foods by low pH, *Mechanisms of Action of Food Preservation Procedures* (G. W. Gould, ed.), Elsevier, London, 1989, pp. 119-160.

42. D. A. Corlett and M. H. Brown, pH and acidity, *Microbial Ecology of Foods* (J. H. Silliker, R. P. Elliot, A. C. Baird-Parker, F. L. Bryan, J. H. B. Christian, D. S. Clark, J. C. Olson, and T. A. Roberts, eds.), Academic Press, New York, 1980, pp. 92-111.

43. T. Eklund, Organic acids and esters, *Mechanisms of Action of Food Preservation Procedures* (G. W. Gould, ed.), Elsevier, London, 1989, pp. 161-200.

44. L. Leistner, Hurdle effect and energy saving, *Food Quality and Nutrition* (W.K. Downey, ed.), Applied Science Publishers, London, 1978, pp. 553-557.

45. D. F. Splittstoesser, Fruit and fruit products, *Food and Beverage Mycology* (L. R. Beuchat, ed.), Van Nostrand Reinhold, New York, 1987, pp. 101-128.

46. S. Sajur, Preconservacin de Duraznos por Métodos Combinados, M.S. Thesis, Universidad Nacional de Mar del Plata, Agentina, 1985.

47. S. M. Alzamora, M. S. Tapia de Daza, A. Argaiz, and J. Welti, Application of combined methods technology in minimally processed fruits. *Food Research International* 26: 125-130 (1993).

48. A. López-Malo, E. Palou, J. Welti, P. Corte, and A. Argaiz, Shelf-stable high moisture papaya minimally processed by combined methods. *Food Research International* 27(6): 545-553 (1994).

49. S. M. Alzamora, L. N. Gerschenson, P. Cerrutti, and A. M. Rojas, shelf-stable pineapples for long-term non refrigerated storage. *Lebensm.-Wiss. u.-Technol.* 22: 233-236 (1989).

50. S. Guerrero, S. M. Alzamora, and L. N. Gerschenson, Development of a Shelf-stable banana purée by combined factors: microbial stability. *J. Food Prot.* 57: 902-907 (1995).

51. S. Notermans, P. in't Veld, T. Wijtzes, and G. C. Mead, A user's guide to microbial challenge testing for ensuring the safety and stability of food products, *Food Microbiology* 10: 145-157 (1993).

52. P. Cerrutti, S. M. Alzamora, and J. Chirife, A multi-parameter approach to con-

trol the growth of *Saccharomyces cerevisiae* in laboratory media, *J. Food Sci.* 55: 837-840 (1990).

53.   M. S. Tapia de Daza and C. E. Aguilar, Effect of combined stress factors in laboratory media, on the growth of *Zygosaccharomyces rouxii* isolated from an I\intermediate moisture papaya product. Paper No. 684, presented at the IFT Annual Meeting, July 10-14, Chicago, Illinois, 1993.

54.   A. López-Malo, L. Parra, and A. Argaìz, Individual and combined effects of water activity, pH and potassium sorbate concentration on the growth of three molds, Paper presented at IFTEC, The Hague, November 15-18, Netherlands (1992).

55.   A. Tablante and M. S. Tapia de Daza, Sobrevivencia de *Zygosaccharomyces rouxii, Saccharomyces cerevisiae, Aspergillus ochraceus* y *Staphylococcus aureus* en lechosa de humedad intermedia. Paper Presented at the II Congreso Latinoamericano de Microbiología de Alimentos, Caracas, November 5-10, Venezuela, 1989.

56.   R. V. Díaz, M. S. Tapia de Daza, G. Montenegro, and I. González, Desarrollo de productos de mango y papaya estabilizados por métodos combinados. *CYTED-D Boletín Internacional de Divulgación-Universidad de Las Américas-Puebla, México* 1: 5-21 (1993).

57.   P. Corte, R. V. Díaz, A. Tablante., A. Argaíz, and A. López-Malo, Ensayos de reto microbiano con *Zygosaccharomyces rouxii* y *Aspergillus ochraceus* en frutas conservadas mediante la tecnología de obstáculos. *CYTED-D Boletín Internacional de Divulgación-Universidad de Las Américas-Puebla, México* 2: 24-29 (1994).

58.   R. L. Shewfelt, Quality of minimally processed fruits and vegetables. *J. Food Quality* 10: 143-156 (1987).

59.   H. K. Leung, Influence of water activity on chemical reactivity, *Water Activity: Theory and Applications to Food* (L. B. Rockland and L. R. Beuchat, eds.), Marcel Dekker, New York, 1987.

60.   T. P. Labuza and M. Slatmarch, The non enzymatic browning reaction as affected by water in foods, *Water Activity: Influences on Food Quality* (L. B. Rockland and G. F. Stewart, eds.), Academic Press, New York, 1981.

61.   A. Argaiz and A. López-Malo, Estabilidad de Durazno Conservado por Factores Combinados Durante el Almacenamiento. Paper presented at II Congreso Latinoamericano y del Caribe de Ciencia y Tecnología de Alimentos—XXIV Congreso Nacional de Ciencia y Tecnología de Alimentos de México; March 14-17, Mexico City (1993).

62.   J. Chirife, C. Ferro Fontan, and E. A. Benmergui, The prediction of water activity in aqueous solutions in connection with intermediate moisture foods. IV. a_w prediction in aqueous non-electrolyte solutions. *J. Food Technol.* 15: 59-70 (1980).

63.   P. Cerrutti, S. L. Resnik, A. Seldes, and C. Ferro Fontan, Kinetics of deteriorative reactions in model food systems of high water activity: glucose loss,

5-hydroxymethylfurfural accumulation and fluorescence development due to nonenzymatic browning. *J. Food Sci.* 50: 627-629 (1985).

64. C. Montes de Oca, L. N. Gerschenson, and S. M. Alzamora, Effect of the Addition of Fruit Juices on Water Activity of Sucrose-Containing Model Systems During Storage. *Lebensm.-Wiss. u.-Technol.* 24: 375-377 (1991).

65. A. Monsalve-González, G. V. Barbosa-Cánovas, and R. P. Cavalieri, Mass transfer and textural changes during processing of apples by combined methods. *J. Food Sci.* 58: 1118-1124 (1993).

66. E. Maltini, D. Torreggiani, B. Rondo Brovetto, and G. Bertolo, Functional properties of reduced moisture fruits as ingredients in food systems. *Food Research International* 26: 413-419 (1993).

67. A. H. Rose and B. J. Pilkington, Sulphite, *Mechanisms of Action of Food Preservation Procedures* (G. W. Gould, ed.), Elsevier, London, 1989, pp. 201-224.

68. D. K. Salunkhe, H. R. Bolin, and N. R. Reddy, *Storage, Processing and Nutritional Quality of Fruits and Vegetables, Vol. I and II,* 2nd ed,. CRC Press, Boca Raton, Florida, 1991.

69. S. D. Holdsworth, Fruits, *Effects of Heating on Foodstuffs* (R. J. Priestley, ed.), Applied Science Publishers, London, 1979, pp. 255-305.

70. G. W. Gould and N. J. Russel. Sulphite. *Food Preservatives* (N. J. Russel and G. W. Gould, eds.), Blackie Academic & Professional, London, 1991, pp. 267-283.

71. A. López-Malo, E. Palou, P. Corte, and A. Argaiz, Storage stability of pineapple slices preserved by combined methods, *1995 IFT Annual Meeting: Book of Abstracts*, 1995, p. 68.

72. A. Argaiz, Alternativas de proceso para frutas tropicales, *Memorias del Simposio Nacional de Fisiología y Tecnologa Postcosecha de Productos Hortícolas en México* (M. Yahia Elhadi and C. Higuera, eds.), Limusa, Mexico, 1991, pp. 153-163.

73. A. Argaiz, A. López-Malo, J. Welti, and S. M. Alzamora, Fruit preservation by combined method. Paper presented at ACS Symposium on Minimally Processed Fruits and Vegetables, August 25-30, New York (1991).

74. A. Argaiz, E. Palou, and A. López-Malo, Estabilidad durante el almacenamiento de papaya conservada por factores combinados aplicando concentracin osmótica y secado, *Avances en Ingeniería Química 1994*, Academia Mexicana de Investigación y Docencia en Ingeniería Química, Mexico, 1994, pp. 21-24.

75. A. Monsalve-González, G. V. Barbosa-Cánovas, R. P. Cavalieri, A. J. McEvily, and R. Iyengar, Control of browning during storage of apple slices preserved by combined Methods. 4-Hexylresorcinol as anti-browning agent. *J. Food Sci.* 58: 797-800, 826 (1993).

76. E. Forni, A. Polesello, and D. Torreggiani, Changes in anthocyanianins in cherries (*Prunus avium*) during osmodehydration, pasteurization and storage. *Food Chemistry* 48: 295-299 (1993).

77. A. López-Malo, P. Corte, J. Welti, and A. Argaiz, Cinética de oscurecimiento en manzana conservada por métodos combinados, *Avances en Ingeniería Química 1991*, Academia Mexicana de Investigación y Docencia en Ingeniería Química, Mexico, 1991, pp. 25-29.

78. S. L. Taylor, N. A. Highley, and R. K. Bush, Sulfites in foods: uses, analytical methods, residues, fate, exposure assessment, metabolism, toxicity, and hypersensitivity. *Adv. Food Res.* 30: 1-76 (1986).

79. T. T. Langdon, Preventing of browning in fresh prepared potatoes without the use of sulfiting agents. *Food Technol.* 41: 64-67 (1987).

80. J. D. Pointing, R. Jackson, and G. J. Watters, Refrigerated apple slices: preservative effects of ascorbic acid, calcium and sulfites. *J. Food Sci.* 37: 434-436 (1972).

81. A. de Poix, M. A. Rouet-Mayer, and J. Philippon, Action combineé des chlorures et de 'lacide ascorbique sur l'inhibition de brunissements enzymatiques d'un broyat de pommes. *Lebensm.-wiss. u.-Technol.* 14: 105-110 (1980).

82. H. R. Bolin and R. J. Steele, Nonenzymatic browning in dried apples during storage, *J. Food Sci.* 52: 1654-1657 (1987).

83. G. M. Sapers, L. Garzarella, and V. Pilizota, Application of browning inhibitors to cut apple and potato by vacuum and pressure infiltration *J. Food Sci.* 55: 1049-1053 (1990).

84. G. M. Sapers and M. A. Ziolkowsky, Comparison of erythorbic and ascorbic acid as inhibitors of enzymatic browning. *J. Food Sci.* 52: 1732-1733, 1747 (1987).

85. F. Pizzocaro, D. Torreggiani,and G. Gilardi, Inhibition of apple polyphenoloxidase (PPO) by ascorbic acid, citric acid and sodium chloride. *J. Food Processing and Preservation* 17: 21-30 (1993).

86. T. P. Labuza, J. H. Lillermo, and P. S. Taoukis, Inhibition of polyphenolxoydase by proteolytic enzymes, Killer enzymes XX Symposium International Federation of Fruit Juice Producers, Scientific Technical Commision, Paris, 1990.

87. A. J. McEvily, R. Iyengar, and W. S. Otwell, Inhibition of enzymatic browning in foods and beverages. *CRC Critical Reviews in Food Science and Nutrition* 32: 253-273 (1992).

88. A. T. Paulson, J. Vanderstoep, and S. W. Porrit, Enzymatic browning of peaches: effect of gibberellic acid and etephon on phenolic compounds and polyphenoloxidase activity. *J. Food Sci.* 45: 341-345 (1980).

89. J. D. Dziezak, Monsantos new product extends produce freshness. *Food Technol.* 42: 98 (1988).

90. G. S. Zaragoza, A. López-Malo, and A. Argáiz, Evaluation of anti-browning mixtures for apple preserved by combined methods, *Proceedings of the Poster Session. International Symposium on the Properties of Water. Practicum II* (A. Argaiz, A. López-Malo, E. Palou, and P. Corte, eds.), Universidad de las Américas—Puebla, México, 1994, pp. 217-220.

91. A. Monsalve-González, G. V. Barbosa-Cánovas, A. McEvily, and R. Iyengar, Inhibition of enzymatic browning in apple products by 4-hexylresorcinol. *Food Technol.* 49(4): 110-118 (1995).

92. M. C. Bourne, Effects of water activity on textural properties of food, *Water Activity: Theory and Applications to Food* (L .B. Rockland and L. R. Beuchat, eds.), Marcel Dekker, New York, 1987.

93. R. Ilker and A. S. Szczesniak, Structural and chemical bases for texture of plant foodstuffs. *J. Texture Studies* 21: 1-36 (1990).

94. S. L. Vidales, M. A. Castro, and S. M. Alzamora, Conservacion de frutillas: influencia del proceso sobre la textura y la pared celular. *Libro de Actas del VI Congreso Argentino de Ciencia y Tecnología de Alimentos y 1er. Encuentro de los Tecnicos de Alimentos del Cono Sur*, April 6-9, Buenos Aires, Argentina, 1994.

95. M. C. Bourne, Effect of water activity on textural profile parameters of apple flesh. *J. Texture Studies* 17: 331 (1986).

96. T. Beveridge and S. E. Weintraub, Effect of blanching pretreatment on color and texture of apple slices at various water activities. *Food Research International* 28: 83-86 (1995).

97. T. Ohlsson, Minimal processing—preservation methods of the future: a review. *Trends in Food Science and Technology* 5(12): 341–344 (1994).

98. C. Nguyen-the and F. Carlin, The microbiology of minimally processed fresh fruits and vegetables. *CRC Critical Reviews in Food Science and Nutrition* 34(4): 371-401 (1994).

99. C. C. Huxsoll and H. R. Bolin, Processing and distribution alternatives for minimally processed fruits and vegetables. *Food Technol* 43(2): 132-135 (1989).

100. L. R. Beuchat, Extraordinary heat resistance of *Talaromyces flavus* and *Neosartorya fisheri* ascospores in fruit products. *J. Food Sci.* 51: 1506-1510 (1986).

101. J. I. Pitt, Resistance of some food yeasts to preservatives. *Food Technol. Australia* 26: 238-241 (1974).

102. J. I. Pitt and K. C. Richardson, Spoilage by preservative-resistant yeasts. *CSIRO Food Research* 33: 80-85 (1973).

103. S. Bills, L. Restaino, and L. Lenovich, Growth response of an osmotolerant sorbate-resistant yeast, *Saccharomyces rouxii,* at different sucrose and sorbate levels. *J. Food Prot.* 45(12): 1120-1124 (1982).

104. L. A. Shelef, Antimicrobial effects of spices. *J. Food Safety* 6: 29-44 (1983).

105. L. R. Beuchat and D. A. Golden, Antimicrobials occurring naturally in foods. *Food Technol.* 3: 134-142 (1989).

106. M. J. Jay and G. M. Rivers, Antimicrobial activity of some food flavoring compounds, *J. Food Safety* 6: 129-139 (1984).

107. R. G. Board and G. W. Gould, Future prospects, *Food Preservatives* (N. J. Russel and G. W. Gould, eds.), Blackie Academic & Professional, London, 1991, pp. 267-283.

108. P. Cerrutti and S. M. Alzamora, Inhibitory effects of vanillin on some food

spoilage yeasts in laboratory media and fruit purées. *Int. J. Food Microbiol.* 29: 379-386 (1996).

109.   A. López-Malo, S. M. Alzamora and A. Argaiz, Effect of natural vanillin on germination time and radial growth rate of moulds in fruit based agar systems. *Food Microbiology* 12: 213-219 (1995).

110.   D. Knorr, New developments in non-thermal food processing, *1995 IFT Annual Meeting: Book of Abstracts*, 1995, p. 187.

111.   D. Torregiani, E. Forni, M. L. Erba, and F. Longoni, Functional properties of pepper osmodehyrated in hydrolyzed cheese whey permeate with or without sorbitol. *Food Research International* 28(2): 161-166 (1995).

112.   A. Argaiz, M. Cossio, and A. López-Malo, Storage stability improvement of apple preserved by combined factors, *1995 IFT Annual Meeting: Book of Abstracts*, 1995, p. 67.

113.   C. A. Alvarez, R. Aguerre, R. Gmez, S. Vidales, S. M. Alzamora, and L. N. Gerschenson, Air dehydration of strawberries: effects of blanching and osmotic pretreatments on the kinetics of moisture transport. *J. Food Engineering* 25: 167-178 (1995).

114.   J. Welti, E. Palou, A. López-Malo, and A. Balseira, Osmotic concentration-drying of mango slices. *Drying Technology* 13(1-2): 405-416 (1995).

115.   P. Fito and A. Chiralt, An update on vacuum osmotic dehydration, *Food Preservation by Moisture Control. Fundamentals and Applications* (G. V. Barbosa-Cánovas and J. Welti-Chanes, eds.), Technomic, Lancaster, PA, 1995, pp. 351-374.

116.   X. Q. Shi, P. Fito, and A. Chiralt, Influence of vacuum treatment on mass transfer during osmotic dehydration of fruits. *Food Research International* 28(5): 445-454 (1995).

Index